Ecological Restoration and Environmental Change

What is a natural habitat? Who can define what is natural when species and ecosystems constantly change over time, with or without human intervention? When a polluted river or degraded landscape is restored from its damaged state, what is the appropriate outcome? With climate change now threatening greater disruption to the stability of ecosystems, how should restoration ecologists respond?

Ecological Restoration and Environmental Change addresses and challenges some of these issues which question the core values of the science and practice of restoration ecology. It analyzes the paradox arising from the desire to produce ecological restorations that fit within an historical ecological context, produce positive environmental benefits and also result in landscapes with social meaning. Traditionally restorationists often felt that by producing restorations that matched historic ecosystems they were following nature's plans and human agency played only a small part in restoration. But the author shows that in reality the process of restoration has always been defined by human choices. He examines the development of restoration practice, especially in North America, Europe and Australia, in order to describe different models of restoration with respect to balancing ecological benefit and cultural value. He develops ways to balance more actively these differing areas of concern while planning restorations.

The book debates in detail how coming global climate change and the development of novel ecosystems will force us to ask new questions about what we mean by good ecological restoration. When the environment is constantly shifting, restoration to maintain biodiversity, local species, and ecosystem functions becomes even more challenging. It is likely that in the future ecological restoration will become a never-ending, continuously evolving process.

Stuart K. Allison is a Professor of Biology and Director of the Green Oaks Field Research Center at Knox College, Galesburg, Illinois, USA.

Ecological Restoration and Environmental Change

Renewing damaged ecosystems

Stuart K. Allison

First published in paperback 2014

First published 2012
by Routledge
4 Park Square, Milton Park, Abingdon, Oxon OX14 4RN

and by Routledge
605 Third Avenue, New York, NY 10017

Routledge is an imprint of the Taylor & Francis Group, an informa business

British Library Cataloguing in Publication Data
A catalogue record for this book is available from the British Library

Library of Congress Cataloging in Publication Data
Allison, Stuart K.
Ecological restoration and environmental change : renewing damaged ecosystems/By Stuart K. Allison.
p. cm.
Includes bibliographical references and index.
1. Restoration ecology. I. Title.
QH541.15.R45A45 2012
639.9–dc23

2012001956

ISBN: 978-1-84971-285-9 (hbk)
ISBN: 978-1-138-80456-2 (pbk)
ISBN: 978-0-203-12803-9 (ebk)

Typeset in Baskerville
by Wearset Ltd, Boldon, Tyne and Wear

To my wife Holly

"It is our choices, Harry, that show what we truly are, far more than our abilities." Spoken by Albus Dumbledore.

J.K. Rowling, *Harry Potter and the Chamber of Secrets*

Contents

Illustrations

Figures

Tables

Acknowledgments

A beginning

My father used to have a small poster in his office that showed a photograph of a young boy dressed like Tom Sawyer, sitting on a log, his elbows propped on his knees, his head supported by both hands. The caption to the poster said "Sometimes I sits and thinks and sometimes I just sits." Writing a book requires lots of time for sitting and thinking and even for just sitting. First I must thank the US-UK Fulbright Commission for providing the funding that allowed me to spend a year sitting, thinking and even writing. Professor Jim Harris and Cranfield University were exemplary hosts during my Fulbright year, providing support, resources and the space in which to work. Simon Medaney was the perfect guide to English culture and customs. I have been fortunate to have an academic home at Knox College, which has provided me with time, funds, Green Oaks as my laboratory and psychological support. President Emeritus Roger Taylor and Dean Lawrence Breitborde at Knox have been especially helpful.

I have benefited from many conversations with wiser, more experienced restorationists, conservationists, environmental philosophers and artists over the years. I would especially like to thank Pauline Drobney, Joan Ehrenfeld, Tony Gant, Stephan Harding, Jim Harris, Chris Helzer, Eric Higgs, Richard Hobbs, William Jordan, Tim Kasser, Bill Kleiman, Shahid Naeem, Christer Nilsson, Kate Rawles, Pete Schramm, Mark Spence, Jac Swart and Tiemo Timmerman. All of them have introduced me to new ways of thinking about ecological restoration and as a result have helped me refine my own thoughts.

My students have been a constant source of inspiration to me. Their sharp, inquisitive minds have forced me to stay intellectually alive simply so I can keep up with them. They also examine problems with fresh insight, which frequently helps me see those problems in new ways. I do not have space to name all of my students, but the following have all worked closely with me and were especially helpful to me. I sincerely thank Chelsea Bagot, Elle Bechtold, Ali Boris, Megan Butler, Kelly Cadigan, Po Chan, Sara DeMaria, Alex Faulkner, Desmond Fortes, Jessie Frank, Ryan Gerlach, Megan Hill, Christine Harris, Jared Jaggers, Dan King, Kim Kreiling, Robert Kurtz, Jennifer Logan, Brian Marienfeld, Mathys Meyer, Brent

Newman, Megan Owens, Quinn Palar, Jen Parker, Bryan Quinn, Michelle Rafacz, Andrew Raridon, Lindsey Roland, Kenny Ruzicka, Scott Sapp, Besa Schweitzer, Julia Sievert, Julian Trachsel, Tracy Vahlkamp, Nicole Voss, Lindsey Will, Abby York, Alicia Young, Emily Young and Creal Zearing. I also thank every other student who ever took one of my classes or went to the field with me. You have given me more than I have given back to you.

Sharon Clayton, my reference librarian, Irene Ponce and everyone at the Knox College library have been extremely helpful as I put together references for this book. Nancy Hall helped me refine the figures in Chapter 6.

Special thanks to Tim Hardwick, my editor at Earthscan Books, Taylor & Francis. He contacted me and suggested that I write this book, thus providing the initial germ of an idea that grew into the final book. He has been extremely helpful at every step along the way. I would also like to thank Ashley Irons and the entire production staff at Taylor & Francis and at Wearset. James Aronson read the entire first draft and provided many excellent suggestions. Tiemo Timmerman provided comments on Chapters 3, 4 and 5.

Chapter 1 is based on a paper I published in *Restoration Ecology*, volume 15, 2007, pages 601–605. Special thanks to John Wiley and Sons Publishers for allowing me to reuse some of that material in Chapter 1. A special thank-you also goes to the Union of Concerned Scientists, who allowed me to reprint one of their figures as Figure 4.1 in this book.

Last, but certainly not least, I must thank my wonderful family – wife Holly and daughters Gillian and Gwendolyn. They have, for better or worse, been with me on this journey every step of the way. I could not have written a word of this book without their love, support and understanding.

1 You can't not choose

In this chapter I will:

- discuss questions that have occurred to me while engaged in ecological restoration;
- provide a brief discussion of the SER definition of ecological restoration;
- address some critiques of ecological restoration;
- explain why human choice is always a part of ecological restoration and, because it can't be avoided, why it should be embraced;
- discuss the restoration of Nachusa Grasslands as an example of interesting choices in restoration.

One of my favorite things about ecological restoration is the physical labor involved in creating and maintaining restorations. Although the physical work can be hard, tiring and taxing for the muscles, it is almost always satisfying. After several hours in the field, whether I'm planting seeds, burning prairies or clearing brush, I can look up, survey the area, observe the changes and see that today I got a lot of work done. Engaging in repetitive physical labor also allows me the freedom of letting my mind wander, and the work is an excellent way for me to unwind from the usual pressures of faculty life. When I'm out in the field I don't have to worry about planning a lecture, grading, committee reports, reviewing a manuscript or any other typical duties. Instead I can open up and just absorb the experience of working in the outdoors. Almost always, once I've been working for a while, I fall into a ruminative state in which my thoughts will turn to things I like – my family, my dog, my students, what is going well at the college, sometimes something as simple as the pure joy of being out in the sun with its warmth filtering through my shirt and easing my muscles. On a really good day in the field my mind moves beyond an uncomplicated reflection on my current joys to a free-ranging consideration of new and creative ideas. My best thinking and ideas almost always arise from a good day working in the open.

But that is on a good day. Not all days in the field are good. One day a couple of years ago I was laboring on a hot, humid July afternoon, the sun not easing my muscles but instead beating down on my head, its rays seeming to bore straight through my skull despite my hat. I was engaged in an on-going, never-ending struggle to remove autumn olive (*Elaeagnus umbellata*) from a restored prairie I manage. I crawled under a sprawling autumn olive shrub made up of dozens of stems sprouting from a single dense rootstock (clearly I had tried and failed to eradicate this individual at some time in the past). The shrub was about 10 feet tall and easily spread out to cover a 12-foot diameter of earth. The branches of the shrub reached out and bent down to the ground so I had to crawl on my belly through a thick barricade of over-hanging branches to reach the actual upright growing stems. I had already been working for many hours, crawling under several similar autumn olive shrubs. I was cutting the shrubs down with a handsaw because I was working alone and I thought lying on the ground, on my side, operating a chainsaw a few inches from my face would be too dangerous and noisy. To protect myself from the hanging branches and short spines I was wearing a thick long-sleeved shirt, heavy overalls, boots and leather gloves. Due to the heat and humidity I was sweating profusely and the sweat had completely soaked through my clothing. As I crawled under the shrubs, soil got caked onto my clothes and face. Wood chips from the sawing were also glued to my face and hair and many had fallen down the back of my shirt. I could feel sweat running in rivers down my body and occasionally felt "things" crawling on me – things I feared would be Lyme disease-infested ticks looking to use me for a blood meal. As I crawled under this particular shrub I noticed a very heavy, sickly sweet, pungent odor, the odor of death and decay. The deeper I crawled under the shrub, the more intense the already strong stink became. Clearly there was something dead under this shrub, but I couldn't see what it was from my vantage point with my head just a few inches above the ground and a thicket of stems in front of me. Well, I thought I was already under the shrub; I might as well cut away the branches and then paint the stumps with herbicide. But with each branch cut and tossed aside, I got deeper into the bush and the smell grew even worse. What on earth was lying there dead? And what did it die of? Did it die of something that I could contract from breathing in the fetid air? How gross was it going to be when I could finally see it? I kept going, cutting and tossing away branches, trying to breathe through my sweat-soaked shirt, which I had pulled up over my nose. Finally I cut away enough branches to see the source of the smell – a dead and decaying raccoon that lacked any obvious injuries. It must have just crawled under the autumn olive to die. Was I thinking of the joys in my life or creatively developing new theories about ecological restoration while cutting down this shrub? Not at all. Instead, the entire time I was sawing away on that shrub, I was thinking to myself (and hearing David Byrne sing a line from an old Talking Heads song), "Well, how did I get here?"

Indeed, whenever I'm up to my eyeballs in some unanticipated problem during ecological restoration I often find myself wondering how I got there and what chain of events led me to be in this particular position at this particular time. When I'm clearing brush to remove invasive species, I frequently ask myself why I'm doing this work. I am a plant ecologist, trained in botanical natural history and I have a deep love of plant diversity in all its forms. I am always amazed by the many adaptations that allow different species of plants to succeed in their environment. Plants amaze me because "they can eat light" (to quote Tim Plowman (Davis 1996, p. 40)), and as long as there is the tiniest amount of soil, light, water and warmth, some plant will be found growing there. So why do I find myself working so hard to eliminate a wonderfully well-adapted plant (the autumn olive) that is incredibly successful (too successful by half) in my restored prairie? Autumn olive grows quickly, more than holds its own in competition with other plants, provides valuable food for birds and mammals with its fruits, and its dense shrubs make excellent wildlife cover. And yet, the more time I spend trying to eliminate it, the more I dislike it. Sometimes I even hate autumn olive – and that seems a very strong overreaction to a perfectly good plant. What is not to like about autumn olive? Much, it turns out. And I am not alone in disliking it here in North America. It is an introduced, invasive plant native to Asia that has been declared a noxious weed by both federal and state agencies in the United States (http://plants.usda.gov/java/profile?symbol=ELUM accessed June 28, 2010), primarily because it spreads so rapidly and out-competes native North American plant species.

The exact chain of events that led to me lying under that particular autumn olive with a dead raccoon for a partner would be long and difficult to reconstruct. The most basic summary is that I grew up in a small town in rural Illinois. During my childhood I spent many hours wandering around town, playing in a nearby pond, vacant lots, small adjacent woods and parks and, as I got older, biking into the countryside to explore other seemingly wild places. My maternal grandfather's farm ten miles north of town was almost a second home for my family – we went there often for celebrations, picnics or just to visit. I also spent many hours walking across his fields and pastures, along hedgerows and down to a stream on the edge of the farm. I was always catching frogs and toads, bringing home tadpoles that I tried to raise (without much success) in an old goldfish bowl, catching butterflies and beetles. When I went off to college my goal was to study biology, especially the natural history end of biology, because I loved exploring the nooks and crannies of the natural world. Eventually I earned a Ph.D. in ecology, studying the role of disturbance in coastal salt marshes in northern California. But when I finished my Ph.D. I was left wondering, "Well, what does that work mean? What is its value?" I had done research that I was proud of, but it was research that I was afraid would only be of interest to a very small circle of specialists working on similar questions in coastal salt marshes (a quick search of citations for my papers on disturbance in salt marshes proves that my fears were well founded). I, like many people, wanted to do scientific research that was of relevance to more than just a few other highly specialized scientists.

Another thing I noticed when I was in graduate school was that ecologists had a very strong preference for studying ecosystems they thought were in some way "pristine," relatively undisturbed by human activity. In fact, human-dominated ecosystems were considered unrepresentative of typical ecosystems; the processes operating in them were thought to be in some way unnatural, and thus studies of human-dominated ecosystems were suspect because the results were likely tainted by human presence. At the time that seemed odd to me given that humans occurred in such large numbers across most of the land. In fact, in coastal California where I was working it was extremely difficult to even find ecosystems that were not influenced by humans. I thought that studies of ecosystems either dominated or disturbed by humans would be of more relevance to most people, especially as we became more and more aware of losses of biodiversity and environmental change brought about by human activity (McKibben 1989). Unfortunately, that preference for studying ecosystems never or seldom disturbed by humans continues to this day (Marris 2009).

I was doing postdoctoral research at Rutgers University, working with Joan Ehrenfeld, when I first learned about ecological restoration and its associated science, restoration ecology. Joan was involved with ecological restoration at the huge Fresh Kills Sanitary Landfill on Staten Island, New York. At first I couldn't believe it was possible to actually rebuild ecosystems on a site as damaged as Fresh Kills. But once I realized that not only was it possible, it was already being done, I thought "here it is" – a way to do excellent ecological research that is also relevant to many people far beyond the small, somewhat cloistered world of those who usually read ecology journals. Although I didn't do any restoration ecology at Rutgers or at Fresh Kills, learning that ecological restoration was happening changed my ideas of what was possible and set me on a new path for my research and career.

What is ecological restoration?

I will discuss the history and definitions of ecological restoration in detail in Chapter 2, but for now I will provide the most widely accepted definition of ecological restoration in order to set the stage for the rest of this book. In 2004 the Society for Ecological Restoration (SER) defined ecological restoration in its *Primer on Ecological Restoration*:

> Ecological restoration is the process of assisting the recovery of an ecosystem that has been degraded, damaged, or destroyed.
>
> (SER Science and Policy Working Group 2004)

The 2004 SER definition is intentionally broad to ensure many types of restoration fit, and does not say much about the process of restoration. In the introduction to the *Primer*, the SER provided a longer statement about the process and goals of ecological restoration:

> Ecological restoration is an intentional activity that initiates or accelerates the recovery of an ecosystem with respect to its health, integrity, and sustainability. Frequently, the ecosystem that requires restoration has been degraded, damaged, transformed or entirely destroyed as the direct or indirect result of human activities.... Restoration attempts to return an ecosystem to its historic trajectory.
>
> (SER Science and Policy Working Group 2004)

There are two key points that emerge from the expanded definition. The first is that ecological restoration is an intentional activity – it is something that restorationists actively choose to do in order to ensure that a damaged ecosystem recovers better ecosystem function (defined as improved health, integrity and sustainability). The second is that restoration "attempts to return an ecosystem to its historic trajectory." Thus, ecological restoration is not simply about improving ecosystem function, but also about producing a restored ecosystem that is similar (in terms of species composition and ecosystem function) to the ecosystem that existed prior to being damaged. Also crucial to the second key point is that we acknowledge that the original, pre-damage ecosystem would have been a dynamic ecosystem, changing in response to both external and internal forces even without the human-caused damage ever occurring. We must restore ecosystems so that they are able to continue to evolve and change as they would have without the human disturbance that necessitated restoration.

Many restorationists engage in ecological restoration thinking that it is an inherently good thing for us to repair human-caused damage to ecosystems. In North America at least, the end product of ecological restoration is often thought to be intuitively obvious as well – we will restore to the conditions that existed prior to the arrival of European-Americans and the introduction of new land-use practices by them. As I began my work at Knox College's Green Oaks Field Research Center, I frequently reflected on the choices made by my predecessors. Back in 1955 Paul Shepherd and George Ward had decided to scatter the seeds of prairie plants to re-establish tallgrass prairie on a former agricultural field. During the 1960s, 1970s and 1980s Pete Schramm continued the restoration efforts of Shepherd and Ward and expanded on them, until eventually there were 40 acres (19 ha) of restored prairie at Green Oaks. When I was hired in 1997 I was specifically instructed that one of my duties was to maintain and continue the prairie restorations. Thus a series of choices that began before I was even born resulted in a set of circumstances that would eventually lead me to spend a hot July day cutting down autumn olive to prevent it from over-growing and out-competing the restored prairie planted so many years before. During my work at Green Oaks I frequently ask myself: Why did they choose to restore prairie here? Why did they select the particular species of plants planted here? I often wonder just what species they chose to plant because the original species list of seeds initially planted has disappeared and Shepherd and Ward are both deceased, so their original flora is now lost to posterity. Why did they plant in such small, irregularly shaped patches with so much wooded perimeter – the perfect kind of long, convoluted edge that favors encroachment by invasive woody plants like autumn olive and black locust (*Robinia pseudoacacia*)? But above all, I think about how choices made at the beginning of restoration in 1955 and subsequent choices since then all influence management of the prairies today and will continue to influence the ecology and management of prairies at Green Oaks long after I am gone.

Today we engage in ecological restoration to produce ecosystems that have health, integrity and that are sustainable, while returning to a historical trajectory that may include dynamic changes to ecosystem structure and function, thus allowing evolutionary and ecological processes to operate. We also have a goal of producing restorations that will be as free from human influence as possible so that they will have self-sustaining health and integrity. But it is obvious that the entire program of ecological restoration is influenced by human choice from its very beginning. Humans decide what ecosystems to restore, what the end goal is, where the restoration will be located and the methods used to achieve restoration. All of those choices have profound effects on the restored ecosystem and will influence/determine the kinds of ecosystems that develop and our interactions with the ecosystem during and after restoration. Based on my own experiences working with tallgrass prairie restoration in Illinois, in the so-called "prairie peninsula," it is difficult for me to imagine a restored ecosystem ever becoming free from active human management. Tallgrass prairies in the prairie peninsula have been maintained by human-set fires for thousands of years, and without Native American use of fire those prairies would not have existed when Europeans arrived in the area (Anderson 2006). Restored tallgrass prairies are completely dependent on fires for their maintenance and renewal (Curtis and Partch 1948; Collins and Wallace 1990) and such fires almost only occur due to human activity today. Thus, at least with tallgrass prairies, human choice and agency in restoration will never end.

In many ways the critical questions facing ecological restorationists in the twenty-first century are: First, how do we reconcile the role of human choice in restoration with our desire to produce self-sustaining ecosystems exhibiting health, integrity and normal structure and function? Second, what role will human choice play as we restorationists wrestle with the twin pressures of global climate change and invasive species? The effects of both climate change and invasive species are a double whammy to the earth that are likely to intensify in the twenty-first century and result in the creation of "novel ecosystems" (Hobbs *et al.* 2006; Seastedt *et al.* 2008). Novel ecosystems are ecosystems that contain mixtures of species that have never before existed on earth due to the spread of invasive species and changes in the distribution of native species as a result of changing climatic conditions at both local and global scales. The combination of global climate change and invasive species will greatly complicate our efforts to preserve and restore ecosystems (Harris *et al.* 2006). Even if I somehow win my seemingly Sisyphean battle with autumn olive, there is likely to be another invasive species waiting in the wings to take its place, and my favored tallgrass prairie species may not be well suited to the conditions on the ground in Illinois if climate change is as extreme as it is predicted to be (Allison 2011).

Critiques of human agency in restoration and some responses

When I first began working in ecological restoration I, like many others, thought ecological restoration was a truly wonderful, inherently good enterprise – something we should all engage in as much as possible. Thus I was shocked when I first stumbled across articles by Robert Elliot (1982) and Eric Katz (1992) that were highly critical of ecological restoration. I did not understand how anyone could fault the effort to restore degraded habitat. Elliot and Katz made the claim that restored habitats were not natural, they were in fact artifacts because only the environment can create natural objects while humans create artificial objects, artifacts that are mere facsimiles of the original model. Steven Vogel has argued that what humans do is only possible because of our evolved abilities (large brains, opposable thumbs, etc.); thus we cannot do anything that is in fact unnatural – or another way of looking at it is that our seeming artifacts are in fact produced by natural processes (our mental and physical adaptations) that evolved over millions of years (Vogel 2003). Although it is somewhat discomforting to do so, we should realize that when clearing forests (for example) the use of bulldozers and herbicides by industrial humans is every bit as natural as the use of hand axes and fire by hunter-gatherers. Indeed, the use of a bulldozer to remove trees is as natural as a beaver using its teeth to cut down trees – it is simply each species using its evolved abilities to accomplish a task.

I do not intend to critique the arguments made by Elliott and Katz because many others have done so already (Higgs 2003; Vogel 2003), but I mention their work to help put the problem of choice in restoration into context. Although I disagree with their overall conclusions, they raise critical questions that restorationists must answer as we consider the meaning and value of our work. One of their main concerns about ecological restoration was centered on the idea that we are better off choosing to preserve whatever wild habitats remain rather than trying to improve already damaged habitat via restoration. I cannot think of anyone working in restoration who would disagree with the need to preserve remaining wild habitat. But what Elliott and Katz fail to acknowledge is that at this stage in human cultural and ecological history the future of all habitats will depend upon human choice, whether our choice is to preserve, restore or continue to develop habitat.

While issuing a call for people to embrace ecological restoration, Frederick Turner wrote:

> We must take responsibility for nature. That ecological modesty which asserts that we are only one species among many, with no special rights, we may now see as an abdication of a trust. *We are, whether we like it or not, the lords of creation*; true humility consists not in pretending that we aren't, but in living up to the trust it implies by service to the greater glory and beauty of the world we have been given to look after.
>
> (Turner 1985, pp. 51–52; emphasis added)

Despite his appeal to our sense of humility, there is an arrogance to his phrase "the lords of creation" and the notion that we have been given the world to oversee that fits in all too well with our history of treating the earth as a resource that is there for us to use as we choose and with a Judeo-Christian tradition of feeling that divine providence delivered all this bounty for our own personal benefit, with no need to consider other species or the earth as a living system (White 1967). This arrogant attitude and resource-use mindset have led to many of the problems that restoration aims to correct. Yet Turner hit upon a fundamental truth – it is clear that given the realities of global climate change, the human imprint on the face of the earth is so large that there is no place that has escaped our influence. If we look for nature and the natural in places without the human footprint, then we will never find them because wild nature as a place devoid of human influence has ceased to exist (McKibben 1989; Rees 2000). Daniel Janzen feels that humans have overwhelmed the environment.

> And all this is to say that humans have won the battle against nature. Humanity makes its living by preventing restoration. It is up to us to accept the responsibility of putting the vanquished back on their feet, paws, and roots. We can do it.
>
> (Janzen 1988, p. 244)

Implicit in Turner's and Janzen's calls for recognition of human domination of nature – and thus duty to restore nature – is the notion that humans are uniquely situated to carry out this restoration. As far as we know we are the only species to have developed moral and ethical systems; thus it is imperative for us to choose to repair the damages we have caused. If we are going to live in a world that continues to possess riches of biodiversity, a world that has room for species and habitats that do not benefit us, then we must restore and preserve as much habitat as possible.

Eric Higgs (2003) sees ecological restoration as a complex discipline in which good restoration must be based on four key points – ecological integrity (the restored ecosystem should function to maintain biodiversity and other key ecological properties); historical fidelity (the restored ecosystem should be based on the ecosystem that was damaged by past human activity); focal practices (restoration should generate meaning for humans through regularly working with nature); and wild design (an acknowledgment that when restoring habitat we are making choices about the future of that habitat but that we must also plan for natural processes taking over and modifying the habitat once our initial efforts at restoration end). Higgs thinks restorationists are key players in determining how nature is defined and used and that they must lead the process of helping restore not just nature but the human relationship to nature (Higgs 2005). Once restorationists fully recognize their role in choosing at least the starting conditions and hoped-for end-point of restoration projects, it is vital for them to engage with both the environment and the human community to achieve the restoration of humans and nature.

Restoration as an active choice to work with the environment

If we heed the call to "accept our responsibility" for repairing the damage we have caused to the non-human world (and I think all restorationists accept that responsibility), we are left with some difficulties. Humans have been choosing how to interact with the environment ever since the first members of *Homo sapiens* walked across the African plains. Our choices are usually made for rational reasons in the short term, reflecting immediate needs and desires, but too often we have not fully considered or understood the long-term consequences of our choices (Hardin 1968). When we realized our actions sometimes have negative consequences for the world we developed two distinct modes of thought about how to respond to our actions, which Evan Eisenberg has termed "Fetishistic" and "Managerial" (Eisenberg 1998). Fetishers are people who think nature knows best and is best left alone by humans as we always muck things up. For a Fetisher the best way to deal with human environmental damage is to set aside preserves where humans never visit or at least leave no trace after visiting. In its most extreme version, a Fetishistic world would be one in which all humans return to a hunter-gatherer lifestyle. On the other hand, the Managers (and restorationists are definitely in the Management camp), think that the only way to repair a damaged earth is for humans to become involved in restoration and management projects designed to reduce human impact and improve the world for other species. Eisenberg, who tends to be a Fetisher, worries that Managers "like managing: they have trouble keeping their hands off" (Eisenberg 1998, p. 287). To a certain extent, he has a point. We restorationists do like working with the environment. It makes us feel better about our relationship with nature if we attempt to correct past human errors. Many restorationists, from Aldo Leopold to Eric Higgs and Bill Jordan, see active human involvement with managing and restoring the environment as key to developing a healthy relationship between humans and the environment (Leopold 1939; Higgs 2003; Jordan 2003). How, then, are we to satisfy our desire to manage and restore nature without mucking it up?

Mark Michael classifies the human impact on nature as interference – some event or action that redirects an ecosystem so that its processes, functioning and structure are different from before the interference (Michael 2001). Because restoration explicitly aims to redirect an ecosystem, it is a form of interference with nature. Restorationists usually intend their restoration of an ecosystem to return the ecosystem to some condition that would be similar to that which existed before an earlier human redirection that we now regard as damaging the ecosystem. Michael concluded that there are degrees of interference, and whether interference is ethically permissible or not must be determined on a case-by-case basis. Michael's conclusion is helpful to us because it allows us room to negotiate whether restoration is proper and identifies that we must make such decisions in light of an examination of our goals and reasons for restoration.

In order for restoration to avoid the problem of being just another managerial interference with nature, for it to really be about repairing the environment and our relationship to the environment, then restoration has to avoid restoring the environment solely to benefit human needs and desires. Restoration must leave room for species that are not beneficial or desirable to humans and to allow the environment to develop along pathways that are not controlled by humans (Gross 2006). Katz continues to see restoration as human domination of nature (Katz 2007), but it need not be so. If restoration is a "co-creative" interaction between humans and the environment, then restoration is not necessarily human domination of nature (Ladkin 2005). Restoration will be an especially co-creative process if we attempt to learn "from the landscape itself, an assumption that the landscape has its own agency and projects" (Ladkin 2005, p. 204). Although assuming that the environment has its own agency is taboo for Darwinian biologists (Davis and Slobodkin 2004), many restorationists at least informally discuss the environment in similar terms, although they may not actually believe in such agency. Aldo Leopold wrote about the land "thinking like a mountain" and discussed the land having particular desires (Leopold 1949). If we think about restoration as an adaptive process with many feedbacks in which both socio-cultural and ecological concerns are addressed, with changes in one influencing the other, while progressing toward a goal of mutual benefit for humans and the environment (Gross 2006), it is possible to envision a co-creative restoration without resorting to assumptions about environmental agency.

Models of implementing human choice in restoration

How, then, do we approach the task of making wise choices in restoration? We usually start by talking about restoration achieving historical fidelity and ecological integrity (Higgs 2003). But there is considerable debate in the restoration community about what those terms mean. Does historical fidelity mean restoring to some past condition and maintaining the ecosystem in that condition, or does it mean restoring to some past condition and then allow the inevitable ecological changes that will occur? Attempting to maintain a restored ecosystem at a particular set of past conditions may result in us creating attractive but ultimately evolutionary sterile and static museum pieces. On the other hand, given the reality of global climate change and the rise of novel ecosystems, allowing for ecological change may result in an ecosystem that bears very little resemblance to the original goal or reference ecosystem. And what exactly is ecological integrity? Is it having a certain amount of biodiversity and particular ecological functions? Does a restoration have ecological integrity if it is missing entire trophic levels? I manage 19 ha of restored tallgrass prairie – an area that is far too small to support bison, the dominant large mammal herbivores characteristic of tallgrass prairies, let alone the large mammal predators that preyed on bison. Does a well-established prairie plant community that lacks large mammals but supports arthropods, small mammals and nesting by rare birds like Henslow's sparrow exhibit ecological integrity? I think it does, but I am biased in this case.

For Eric Higgs and William Jordan the choices in restoration are really about humans choosing to restore both nature and the human relationship to nature. They focus on the process of restoration, seeing the process as an act that reunites humans with nature via regular performance. Higgs describes restoration as a focal practice that generates meaning to humans via a synthesis of mind, body, thing (such as a tool) and the environment. Higgs employs the metaphor of restoration as a conversation, fully cognizant of the fact that nature cannot speak to us in words. He feels that if we approach nature as an equal we should be able to engage in a conversation with nature and decipher the subtle messages nature provides (Higgs 2003). Jordan sees restoration as a ritual that humans use as a way to address the sense of shame that arises due to our own inadequacies in how we treat nature. Jordan sees communion as the primary metaphor for restoration, using communion in both the sense of sharing and spiritual fellowship (Jordan 2003). Although these metaphors are helpful as we envision how to restore nature, ultimately we need to make specific choices for each restoration project and we must move beyond metaphor to implementation.

Given the realities of global climate change and human use of the earth's resources (Vitousek *et al.* 1997), many people want to create new ways for humans to coexist with the rest of the world's organisms. Some scientists and restorationists have developed ideas for creating forms of restoration that they label "win–win ecology," "reconciliation ecology" and "futuristic restoration," the key features of which are planning for human development so that we share habitat and resources with other species, even those not useful or beneficial to us, and planning for restored sites to undergo dynamic changes as the earth's climate continues to change (Rosenzweig 2003; Choi 2004). The goals of reconciliation ecology seem to be a bit too optimistic and human-centered, as the underlying assumption is that humans will continue to develop the earth but will do so in a way that satisfies human needs for resources and beauty while also satisfying the needs of other species. But at this juncture in the relationship of humans to the environment, we need to be more optimistic than not.

The centrality of human choice in restoration forces us to make difficult decisions when choosing our methods and goals in restoration projects. The questions about choice in restoration are almost never black and white – instead they are a gray area. Environmental historian Kenneth Olwig has argued that if we remove human values from nature, then nature will not mean much to us and we will fail to properly appreciate and love it. As we shall see in many examples of restoration projects, love or a deep appreciation of the land is the key starting point for most restorationists. Olwig locates the origin of human values for nature in human interactions with nature and sees the need for recognition of cultural landscapes as the source of human value for nature (Olwig 1995). The choices we make about restoration may place value on wild nature, cultural landscapes or both, but we must choose what makes sense for each situation.

In North America there has been a tendency for restorationists to avoid engaging with cultural choices in ecological restoration. This avoidance is at least partly due to a focus on restoring ecosystems to conditions that existed before European arrival and settlement and a failure to recognize the profound effects First Nations people had on the continent (Denevan 1992). As a result, North American restorationists are far behind their European colleagues in thinking about how to combine the natural and cultural in ecological restoration. Europeans have thought more about these issues partly because in Europe it is harder to define a point in time when there was a drastic change in land-use practices and partly because in Europe there is very little land that is not currently in cultural use, thus making it critical for restoration to incorporate human needs and desires. Dartmoor National Park in England was preserved as a national park because of its grassy, windswept and open appearance. However, when humans first settled the area around 7,000 years ago, the land was covered by an oak forest. It was turned into moors by the use of fire and the presence of domesticated grazing animals. Recent changes in fire regime and grazing have led to an encroachment of woody vegetation on the moors. Should the moors be restored to their grassy, open condition or restored to the pre-human-arrival forests? This is not an easy question to answer, but the open grassy moor is highly valued and thus maintenance of that ecosystem has been the goal of restoration there (S. Harding, Schumacher College, Totnes, Devon, UK, personal communication, 2005). Similarly, in the Netherlands restoration is an extremely complex question because restoration to "wild" conditions often means allowing natural processes to operate in former agricultural lands where the agricultural land itself was created by diking and filling shallow marine habitats – the landscape itself is cultural in origin. Restoration to a condition of pre-human involvement would mean a return to shallow seas and a loss of land – an unacceptable restoration for most Dutch people. The Dutch restoration movement must confront a series of questions about "what kind of nature it really wants, what nature is or could be, and how the relationship between nature, development, and politics should be conceived of" (van der Heijden 2005, p. 428). Dutch restorationists realized that good restoration means local involvement in the decision-making and restoration process and that it is critical to provide a mixture of habitats that promote biodiversity, wild nature and culturally meaningful landscapes (van der Heijden 2005). The European experience with restoration indicates that good restoration will only happen when human choice is clearly acknowledged and incorporated into the process along with goals of increasing biodiversity and ecological integrity of the restored habitat.

We can conclude that good ecological restoration should produce:

1. an ecosystem with historical fidelity to the pre-disturbance ecosystem;
2. an ecosystem that has ecological integrity – i.e., ecological health, sustainability and all its ecological components and functions;
3. an ecosystem free to realize its evolutionary potential;
4. opportunities for continuing human interactions with the habitat – it should help restore the human–nature connection.

The first three points are often summarized as producing an ecological restoration that creates a more natural ecosystem than the damaged/degraded ecosystem that existed prior to restoration.

Case study: Nachusa Grasslands – a case in point

Figure 1.1 The entrance sign for the Nachusa Grasslands, a 3,000 acre (1,214 ha) prairie and savanna restoration site owned and managed by the Nature Conservancy. Nachusa Grasslands are located just to the east of Dixon, Illinois.

Nachusa Grasslands is a prairie restoration site owned by the Nature Conservancy. It is located in north-central Illinois, just east of the town of Dixon. The central area of the Nachusa Grasslands has many limestone outcrops and fairly steep hillsides with shallow soil, making it rather poor land for typical Midwestern row-crop agriculture. Therefore much of the site was never plowed. Instead, the Euro-American settlers in the area used the land mainly for grazing cattle, so the prairies were maintained as pastures. In 1986 the Nature Conservancy recognized that the Nachusa area represented the best location (and in many ways the last chance) for preserving any relatively large areas of native tallgrass prairie left in the state of Illinois. In that year the Nature Conservancy purchased its first land there (250 acres (101 ha)) and developed plans to both maintain the existing prairie remnants and also to restore prairies on intervening and surrounding agricultural land to increase the size of the prairie on the site. Today the Nature Conservancy owns about 3,000 acres (1,214 ha) and manages another 725 acres (293 ha) via easements. Most of the land is tallgrass prairie, but there is also a considerable area of Midwestern oak savanna. The grasslands support at least 600 species of plants and 180 species of birds. There are many rare species of animals and plants there, including Blanding's turtle, Henslow's sparrow, dickcissels, grasshopper sparrows, Hill's thistle (*Cirsium hillii*), fringe-petaled puccoon (*Lithospermum incisium*) and creamy castilleja (*Castilleja sessiliflora*) (www.nature.org/

wherewework/northamerica/states/illinois/preserves/art1116.html accessed July 5, 2010).

To me, Nachusa Grasslands is one of the most successful and unique restoration projects in North America because of the role of human choice in its planning and restoration – choice that originates with the project director, Bill Kleiman. Bill has been at Nachusa since 1993, when he was hired as its first full-time director. Even today Nachusa only has a full-time paid staff of two people – Bill and an assistant. Bill realized very early in his tenure at Nachusa that he could not possibly restore or manage the site with just two people. He knew he would have to depend on volunteers and some summer interns to get all the work done, but by far the most important contributors to the restoration and management of Nachusa would be the volunteers. The need to use volunteer labor is true at many restoration projects, but Bill has succeeded in getting volunteers to work so hard and so well at the restoration due to the combination of freedom and responsibility he has given them. Bill realized that people are more likely to become committed volunteers, willing to work many hours in the field for free, if they feel like they have ownership of the restoration process. So Bill divided Nachusa into many small management units – usually just a few acres – which he assigned to specific volunteers. Once a volunteer agreed to be responsible for a particular unit such as Dot's Prairie, that volunteer would have almost complete freedom to decide how and when to manage it. The volunteer would decide which invasive species to remove and how and when to remove them. If the unit was scheduled to be restored, the volunteer would decide what mix of species to plant, when to plant them and how to manage the site to ensure that the plants became well established. The volunteer was also responsible for collecting the seed used in the restoration, so that from the beginning of selecting and harvesting the seed, that person was invested in the restoration process. There are some guidelines about how to manage the prairie units and what might work best at a particular restoration site, but within those fairly broad guidelines, the restorationists have almost complete freedom to work as they think is best for the site. Some of the volunteers joked with me that they tell Bill what needs to be done and what the guidelines should be, and then he just agrees with them.

Bill's use of volunteers is visionary, a term he would probably find a bit embarrassing because he is a soft-spoken, modest and self-effacing person. In May 2010 I took one of my classes to visit Nachusa so they could see the restoration for themselves and learn from master restorationists how much hard work and love is necessary to achieve successful restorations. Bill declined to lead us on the tour, saying he was bad at giving tours. Instead he had two of his volunteers – Jay Stacy and John Schmadeke – lead us on the tour. Jay Stacy is a bit of a legend among prairie restorationists in Illinois, but I had never met him so I was excited to have him lead the tour.

Jay and John are both originally from the Chicago area and both are now retired so they are able to devote many, many hours each month to the ongoing restoration at Nachusa. Jay lives only a few miles from Nachusa and has been working there almost constantly for the past 17 years – he must have retired early because he doesn't look to be past his fifties. I asked Jay how he got involved with restoring prairies at Nachusa. He said he used to work in retail in Chicago and was an avid bird-watcher in his spare time. He heard that Nachusa had nice birds and he particularly wanted to see a grasshopper sparrow. So one day in 1993 he drove down to Nachusa and before he even got out of his car he saw a grasshopper sparrow sitting on the Nachusa Grassland entrance sign. He stopped and watched the sparrow for a few minutes, still sitting in his car, and then saw an upland plover fly by – another bird he was eager to see. He thought "Wow this is a really nice place, they must have other nice things besides birds," and started coming down to Nachusa whenever he had free time to learn more. For several years he lived in a trailer on the Nachusa property until he found a place to live in nearby Oregon, Illinois.

Jay is a legend among prairie restorationists because he is absolutely obsessed with achieving high species diversity in the restoration units he plants. Many prairie restorationists will plant a few species that they know work well. My predecessor at Green Oaks, Pete Schramm, usually relied on 48 species in his restorations (Schramm 1992; Allison 2002). For many prairie restorationists, getting 50 species of plants established would be considered a very successful restoration. But Jay looked at remnant prairies with several hundred native plant species and felt restorationists were setting the bar far too low. He thought that if we wanted to achieve high-diversity prairies, we needed to plant high-diversity seed mixes from the very beginning. Jay also just likes plants and plant diversity. If remnant prairies had high diversity, then he would choose them as models, partly to attempt to match them and partly to satisfy his own liking for lots of diversity.

In order to have a high-diversity seed mix, Jay needed to collect a high diversity of seeds. Jay is a true fanatic when it comes to collecting seeds. He will go wherever he has to in order to find seeds from rare species. And he will spend hours collecting seed every day, bent over or on hands and knees, often under very hot, humid conditions in mid-summer, to ensure he has enough seeds and enough diversity to plant his restoration units. During our tour Jay took us to four prairie units, each about ten acres (4 ha) in size that he planted with seeds from 175 species of prairie plants in 2006, 2007, 2008 and 2009, respectively. Some plants take a while to show up in a restoration after it is planted, but by 2010 he had found 134 plant species growing in his 2006 planting. His success rate is absolutely astounding.

But for my class, even more important than his success at planting high-diversity prairie, the most important lesson Jay taught was the need to be observant and to love what you are trying to restore. Jay's face and hands are weathered from spending years in the outdoors in all kinds of weather at Nachusa – winter's raw cold wind or the searing sun of mid-July, years gathering seed, planting prairies and removing invasive species to help maintain his plantings and the remnants at Nachusa. His clothes are worn, a bit threadbare, tattered and patched in places. Appearance is clearly secondary to restoration for Jay. As we walked the prairies, he was constantly getting down on his hands and knees, crawling along, admiring the plants in flower.

He stopped and gently cupped a fringe-petaled puccoon in his hand and exclaimed, "Oh what a cutie!" A few steps further he saw a creamy castilleja and smiled while he caressed its leaves and flowers: "Look at this little beauty."

Even the more common species – bird's foot violet (*Viola pedata*) or blue-eyed grass (*Sisyrinichium albidum*) – was reason to stop, look, gently touch and murmur "What a beauty." Every new flowering plant we found was an occasion to pause, get down to its level and marvel at its delicate beauty. John Schmadeke said that Jay had mellowed over the years – there was a time Jay was so protective of his little darlings, of his newly germinated seedlings, that he would not let other volunteers walk across his prairie plantings.

As we were leaving, Jay told my students that the future of restoration belonged to them. He thought that restoration today is like "cookery" – we put together things by feel and we can achieve good results, rather like a great chef seasoning to taste. But Jay thought the students would be the ones to determine the numbers so that restoration would be a true science, with exact formulas to achieve great restorations. I thought about that for a moment and said I wasn't sure if we would ever get to that point. I said that restoration will always have an element of art to it, some things will always have to be done by feel and that is why someone like Jay himself, with a unique feel and devotion to place was necessary for restoration. Jay smiled a bit and whispered "Thank you for that."

The volunteers at Nachusa work well together, with Bill Kleiman and with the land to achieve their excellent restorations. There is tremendous camaraderie among them. But they are also a bit competitive too. There is a spirit of "If Ann has 50 species in her restoration unit, then I want 60 in mine." (Jay has raised the bar so high no one tries to compete with him for number of species.) There is also an element of "Wow Mary's restoration is so beautiful with all of the creamy indigo blooming now, I wish mine had some more flowers." The competitiveness also plays out in terms of hours in the field – who spent the most time collecting seeds? Who removed more brush from the savanna restoration? But it is a joyful, helpful competitiveness because every volunteer will pitch in to help any other with a question or a problem. The true genius of Bill's granting freedom, responsibility and ownership to his volunteers is not just that they are willing to work so hard for no monetary payment, but that having so many volunteers each managing small units results in a huge increase in diversity of the plantings and ultimately of the overall Nachusa restoration itself. If one person designed all of the restorations over the 3,000 acres at Nachusa, the restorations would probably have a lot of uniformity, with some variety due to differences in soil type, microclimate, topography and past history. But because each volunteer has a unique, personal vision of what the restoration should look like – some favor high species diversity, some favor forbs or particular forbs, some favor the grasses or particular grasses – the restoration at Nachusa has created a patchwork with tremendous species and landscape diversity. My guess is that the diversity will help buffer Nachusa from the effects of climate change and invasive species. It certainly makes Nachusa a highly dynamic, visually and ecologically interesting site to visit.

Conclusion

When I was an undergraduate, one of my professors, Francis Cousens, told us over and over again, "You can't not choose. Failure to make a choice is a choice." Professor Cousens was talking about moral and political choices we would have to make and meant that if we did not make a carefully considered choice, we were voting for maintenance of the status quo. Restorationists are faced with the same dilemma with respect to ecological restoration. Ecological restoration is similar to conservation biology because both are crisis disciplines – disciplines in which we often have to make decisions and choices before we have enough information to be certain that we are making the best choice, but in which we also know that failure to make a choice will result in further degradation to the environment and loss of things we value in nature (Soulé 1986).

In many ways, the fact that we must struggle with making choices and whether we are making the right choices is good for us. Trying to determine the significance of our actions forces us to confront our assumptions and attitudes about how humans should relate to nature and whether our actions fit our beliefs about that relationship. The role of human choice in restorations opens the doors to both human arrogance about our abilities to pick the right goals and methods for restoration and also human foolishness and error if we make poor choices for our goals and methods. Yet if humans are to work in restoration, we cannot get away from human choice as one of the prime factors in restorations. We know from history that people of good will sometimes make choices that turn out to be mistakes in our interactions with the environment. In the late 1800s a policy of fire suppression made perfect sense to forest ecologists in North America. After all, their job was to protect forests and fire killed trees, resulting in the creation of open meadows. One hundred years later we realized fire suppression was bad for our forests and that fire was a vital part of forest dynamics (Pyne 1982). We must hope that our choices at the least do not make things worse than they are now.

To move forward we must embrace the role of human choice in ecological restoration, promote the benefits of greater interaction between humans and the rest of the world via restoration, and make sure our restorations preserve as much species and habitat diversity as possible. We must be cognizant of the complications that will arise as global climate change and species invasions alter all of our ecosystems – wild, restored and human-dominated – and we must be proactive in our planning and management of restorations so that those complications don't overwhelm us. If we can achieve those things on a larger and larger scale, then we will have done much to promote the survival of many species and habitats that are both ecologically and culturally valuable.

I will explore those issues in the rest of this book. In particular I want to ask: What are our assumptions when we plan restorations? What are our goals during restoration? Why do restorationists make the choices they do when conducting a restoration? How are restorationists planning responses to global climate change and increasing species invasions? Is there consensus about assumptions, goals, choice and potential for restoration in the twenty-first century, or is there a great variety of opinions among restorationists with respect to these issues? The answers to these questions will be complex, but it is critical to remember that ultimately the future success of restoration rests upon the Jay Stacys of the world – dedicated individuals who love a particular landscape and are willing to work long hours in pursuit of a sometimes elusive vision of the restored land. The future of restoration will depend on people who do the good, hard work of restoration, not in pursuit of recognition or riches, but because they feel a connection to the land and a need to repair the damages to that land and because for them ecological restoration is the only choice that makes sense given the current condition of the earth.

2 How did we get here?

A brief history of ecological restoration

In this chapter I will:

- provide an in-depth examination of definitions of ecological restoration;
- briefly describe the history of ecological restoration;
- discuss the growth of ecological restoration as an academic/scientific enterprise, focusing on the field developing in phases;
- conclude by identifying the third, currently evolving phase and a possible way forward for third-phase restorationists.

Definitions of ecological restoration

Despite, or perhaps because, ecological restoration is an endeavor whose diverse practitioners have only recently recognized they are engaged in a set of common activities, there have been many definitions of ecological restoration. The Society for Ecological Restoration (SER) was founded in 1988 and that group has attempted to claim the mantle of being responsible for defining the field, but both prior to the SER's organization and afterwards, many different people have produced definitions of the field. I will briefly examine the breadth of those definitions to set the stage for a discussion of the history of ecological restoration. If we don't agree about what ecological restoration is, it is hard to look to the past to examine how the field formed.

In the middle of the twentieth century there were increasing numbers of people engaged in ecological restoration. Most of them assumed that when they talked about restoration they were all talking about the same thing. The most common starting point for defining ecological restoration was to simply begin with a dictionary definition of restoration such as can be found in the *Oxford English Dictionary* ("Restoration: **1.** The action of restoring to a former state or position; the fact of being restored or reinstated. Also const. *to*. **a.** Of persons. **b.** Of territory, conditions, or things.") and then to apply that definition to an ecological situation. For most people the process of ecological restoration begins when there was some area or ecosystem that had been changed from a previous condition, almost always due to human activity, and then restorationists attempt to bring it back to its original condition – rather like restoring a damaged house or an old automobile or even the frescos in the Sistine Chapel. But ecological restoration has proven to be a much more complex thing than restoring a house or a car or a great work of art. And the complications are tied up in how different individuals define restoration, the initial starting point for restoration and the processes used to achieve restoration.

As I noted in Chapter 1, the most commonly used definition of ecological restoration is the one developed by the SER. When the SER was founded in 1988, one of the first actions of the society was to develop a standard definition of ecological restoration. The SER published a definition in 1990, a revised definition in 1996 and a final (for now – I know from my work as a member of the SER Board of Directors that there are plans to revise and update the definition) revised definition in 2002 (Higgs 2003). That definition and further clarification provided in the *SER Primer on Ecological Restoration* (2004) (both given in Chapter 1) help us to understand just what ecological restoration is about and what most of its practitioners hope to accomplish. That restoration is an intentional activity is a key concept to the definition. But the SER definition was preceded by earlier definitions developed by different people with sometimes different ideas about what restoration is.

Anthony Bradshaw provided some of the most insightful definitions of ecological restoration (1984). He noted that when people carry out restoration they are facilitating a normal ecological process. If an ecosystem is disturbed or damaged in some way (such as by fire, flood, volcanic eruption, mining, logging, urbanization, etc.), it will usually enter into a process of ecological succession and eventually return to either the original condition or perhaps to a different condition than before. If we allow nature to take its course and watch as an ecosystem recovers from damage via succession, Bradshaw said we might call that restoration by neglect. And rather like neglect of anything – children, a garden, a house – there is a chance that the neglected will turn out just fine (the ecosystem returns to its original condition) or it may not be so fine, and in the case of damaged ecosystems, the effects of restoration by neglect are often not desirable by humans. So we engage in the act of restoration to facilitate the process of succession – partly to speed up the process as natural successions can take hundreds of years to return to the original condition and partly to ensure that the end of the restoration process is an ecosystem that is similar to the original and that functions in a way that we think is beneficial to the environment as a whole. Bradshaw wanted to reserve the term restoration for situations in which the goal was specifically to return the damaged ecosystem to its original condition. He spent most of his restoration career working with areas in the United Kingdom damaged by mining and he found that many of those areas were so damaged, usually due to destruction of the soil and hydrology, that it was impossible to return to the original ecosystem. In those situations, Bradshaw felt it was best to repair the damage so that the restored ecosystem functioned better than it would if the damaged mine site was simply abandoned – which frequently resulted in an area of bare ground with high rates of erosion. In such situations he advocated a process he called "replacement," in which the damaged site was repaired so that it supported a viable, functioning ecosystem, full of life, with all its trophic levels and diversity, but where the ecosystem was something that could be supported on the degraded site even if it was not similar to the original ecosystem. In later papers Bradshaw (2002) sometimes referred to this kind of restoration as "reclamation" rather than replacement.

Ecological restoration was first widely popularized in the United States by tallgrass prairie enthusiasts who embraced restoration as a way to increase the rapidly dwindling acreage of tallgrass prairie scattered across the Midwest. In the 1920s they recognized that there was so little prairie left, that the only way to ensure prairie remained viable or even present as an ecosystem was to plant tallgrass prairies wherever and whenever possible. These prairie restorationists liked to distinguish between restoration and what they called reconstruction (Kurtz 2001). They felt restoration was a word that should only be applied if a remnant of the original ecosystem was still in place. Again, the analogy would be to a house or car – if the structure was still present they could restore it. But in many cases they were planting prairie in fields that had been in row-crop agriculture for many years and no longer had any plants or animals from the original prairie ecosystem. In this case they felt that what they were engaged in was actually reconstruction – starting from scratch to build a facsimile or copy of the original. It was as if a house had burned down to the ground, leaving nothing but ashes. In that case it would be impossible to restore what was left, but someone could rebuild or reconstruct the house that stood on that site if a set of plans or drawings of the original existed. Perhaps because I mostly work with prairie restoration in the American Midwest and I was taught by those early prairie enthusiasts, the distinction between restoration and reconstruction makes sense to me and I think it is unfortunate that the terminology has not come into common use. But for most restorationists around the world, the term applied is restoration, whether starting with at least a few remnants of the original ecosystem still in place or starting from zero, with nothing of the original in existence.

It is also critical when discussing restoration to realize that restoration is repair that is conducted following some kind of human damage to an ecosystem. While natural processes like fire, floods, volcanic eruptions, windstorms and avalanches may damage and degrade ecosystems, most members of the SER do not see a need to repair such damage. This is an interesting twist in the historical definitions of restoration, because as I will show when discussing the development of restoration, there have been times in the past when restoration was understood to be an activity to repair damages caused by natural processes. In the past in places like Italy, the natural world was thought to be in decline and only human agency could prevent the eventual collapse of nature (Hall 2005). Even though today there may be very good reasons to repair damage caused by natural processes, such as planting species that will grow quickly and whose roots will help consolidate the soil to prevent further erosion following a landslide, that activity usually is not understood as restoration. Sometimes human attempts to improve a naturally damaged ecosystem are seen as "enhancement," an activity that is usually intended to provide some benefit to humans or ecosystems of interest to humans (Bradshaw 2002).

Restoration ecology is the final term that needs to be defined. William Jordan coined the term restoration ecology in the late 1970s to capture the idea that the experimental nature of restoration was leading to the development of a new scientific discipline rooted in classical ecology, but dedicated to understanding how to repair damaged ecosystems (Jordan 2003). Tony Bradshaw elaborated upon this idea by claiming that ecological restoration was "the acid test" for the field of ecology – if we understood ecosystems well enough to be able to restore them, then we could finally claim that ecology had come of age. If we could actually do good restorations, then that would be conclusive evidence that we know how ecosystems are put together, know the structures that are important for their proper functioning and that we can make predictions about which combinations of species and processes are most likely to persist in nature (Bradshaw 1987). Many restorationists use the term restoration ecology to refer to the scientific study of the process of ecological restoration. Ecological restoration is seen as a broader category of activities, taken from many fields and perspectives (scientific, political, ethical, economic and social), that constitute the actual practice of restoring ecosystems (Higgs 2003).

I will begin examining the history of ecological restoration by discussing the early beginnings of the field before it had coalesced into ecological restoration as we know it today. I will end the chapter by examining how the current field of ecological restoration came together via a series of phases. Phase I can be divided into two parallel tracks, with Track I being the pioneering work of prairie restorationists in the American Midwest, who produced an active community focused on restoration to previously existing, historical conditions. The second track developed from a focus on mine reclamation, in which restorationists took a more pragmatic approach in which improving ecosystem services on severely degraded land was the primary goal. Phase I ended when those two tracks came together to produce what we now think of as ecological restoration. Phase II developed as the result of increasing citizen involvement in ecological restoration and an expansion beyond restoration as the province of academics and engineers. We are currently entering Phase III, in which global climate change is forcing us to rethink our approaches to ecological restoration and even to reconsider our goals when conducting restoration projects.

The early, almost pre-history of restoration

When I first started working in the restoration of tallgrass prairies, the received wisdom among prairie restorationists was that ecological restoration had been invented at the University of Wisconsin Arboretum in the 1930s. Usually Aldo Leopold was credited with starting things by giving a speech at the dedication ceremony for the Arboretum in which he suggested that the Arboretum should contain examples of all the ecosystems that existed at the time of Euro-American settlement of the area. Then Norman Fassett and Ted Sperry, who became the original restoration ecologists in this version of history, initiated restoration of tallgrass prairie with the assistance of CCC workers during the Great Depression. The plantings they planned and supervised beginning in 1936 would eventually become known as the Curtis Prairie. Henry Greene, also working at the Arboretum, began in 1943 to single-handedly plant another tallgrass prairie restoration on a sandy site that eventually came to be named the Greene Prairie. The seeds planted by Fassett, Sperry and Greene germinated, flourished and spread, leading to the rise of ecological restoration and the development of restoration ecology as a field of study, seeds that quickly spread far beyond the American Midwest to the entire globe, leading to a revolutionary change in how humans relate to the world around them.

It is a great story and as a native Midwesterner myself, I found it very appealing. Far too often we Midwesterners feel like our home is treated as a boring piece of flyover country where nothing interesting ever happens. To think that a group of Midwesterners like Leopold, Fassett, Sperry and Greene, working at the quintessential Midwestern land grant university, developed a set of ideas that revolutionized how we think about and respond to the world was simply amazing to me, fantastic in every sense of the word. Prairie restorationists feel a great amount of pride in continuing a tradition that began with the origin of ecological restoration. It is such a great story that we want to believe it is true, and parts of the story are true. But the story is far more complex than the received wisdom I originally learned.

The story of the founding of ecological restoration that I learned was in many ways similar to the story of the founding of the modern science of biology. Just as we know that the modern science of biology was born when Darwin published *On the Origin of Species* in 1859, we also know there were ideas about evolution floating around before Darwin and Wallace independently came up with the mechanism of evolution by natural selection. So it seemed reasonable that ideas about restoration must have been floating around prior to the "Ah-ha" moment in Wisconsin. But even the knowledge that there must have been ideas about restoration floating around prior to 1936 does not do justice to the long history of ecological restoration.

The history of restoration is much like the history of ecology itself – it is an ancient activity (the first ecologists were really hunter-gatherers who had to be excellent natural historians in order to find food and be able to predict when some foods would be available to them (Krebs 2008)) – that also has a long scholarly tradition that coalesced into a field of study in a fairly short time following a long gestation period. We can trace a scholarly tradition in ecology that stretches back at least to Aristotle and Theophrastus, with numerous early biologists and natural historians working throughout the Renaissance. But the exact origins of the modern science of ecology are hard to pinpoint. I could (and usually do) argue that ecology began in 1859 with the publication of *On the Origin of Species* because the theory of evolution by natural selection is a profoundly ecological theory because the environment (both physical and biological) acts as the selective agent on individuals. Or I could claim that ecology began in 1866, when Ernst Haeckel first coined the term. Perhaps ecology began when Wallace published *The Geographical Distribution of Animals*, a monumental synthesis of biogeography in 1876. Eugen Warming is sometimes considered the founder of ecology because of his pioneering work in Brazil, and his book *Plantesamfund* (*Oecology of Plants* in English), published in 1895, is credited with being the first true ecology text. Ecology in America is often thought to have begun with Henry Cowles' study of succession on Lake Michigan sand dunes – work first published in 1899. Ecological restoration has a similarly long and varied history.

Ecological restoration, when defined as intentional human repair of ecosystems damaged by humans, originated long before recorded human history. Hunter-gatherers today practice sophisticated land management and there is excellent archeological evidence that such practices have been conducted for millennia (Vale 2002; Anderson 2005; Jackson and Hobbs 2009). Much of their land management is directed towards making hunting easier, maintaining and promoting the growth of desirable plants and clearing areas for villages and trails. At least some of this management is directed at repairing areas that have declined in terms of providing natural resources desired by hunter-gatherers, and in that sense count as ecological restoration. After the development of agriculture, villagers and small farmers continued traditional land management practices and initiated new ones with the goal of maintaining desirable ecosystem functions and restoring degraded functions. Often the restoration efforts were fairly simple, such as planting seeds to maintain a supply of trees for lumber and firewood and also planting trees, shrubs, and grasses to reduce erosion (Whited 1996; Hall 2005). As more complex societies arose, the need to preserve and restore natural resources such as forests became more formally codified. In Japan, Shinto priests have been supervising the management of forests near their temples in order to maintain a steady supply of timber for at least 2,000 years. Some pages of Virgil's *Georgics* (written in 30 BCE) can be read as a description of the need for everyone living in the countryside to care for trees and maintain forests as an early form of restoration (Dudley *et al.* 2005).

Restoration as an organized, officially sanctioned activity is of more recent origin. The first attempts at forest restoration appear to have been conducted at Nurnberger Reichswald in the fourteenth century (Fischer and Fischer 2006). The Englishman John Evelyn published a pamphlet in the early 1600s about the need to plant trees in order to re-establish English forests (Dudley *et al.* 2005). Hans Carl von Carlowitz wrote the first book on silviculture in 1713. The book was based on his experiences managing timber production in Freiburg/Saxony, and in it von Carlowitz attempted to establish principles by which foresters could properly manage forests in order to achieve sustainable forestry – the first use of the term sustainable (Fischer and Fischer 2006). By the middle of eighteenth century, forest management programs had become established in many parts of Europe (Fischer and Fischer 2006). At the beginning of the nineteenth century local, regional and even national government agencies in France and Italy initiated restoration efforts, using the word restoration to describe their efforts to prevent erosion and limit damage by flood waters during the spring melt and heavy rain storms (Whited 1996; Hall 2005). Marcus Hall found that, in Italy in particular, the early restoration efforts began because increasing floods and erosion in the mountains were thought to be due to a decline in forests that was caused by a lack of human care for the forests. Many Italian scientists and politicians believed that nature was in a constant state of decline and that careful human tending of the natural world was required to halt the decline due to natural processes and to return nature to a more benevolent and productive state. Italians felt that they lived in a vast garden, encompassing their entire peninsula, and that it was their duty to care for that garden. It was not until the late nineteenth and early twentieth centuries that Italian restorationists began to place the blame for erosion and floods on human causes, such as excess logging and grazing high in the mountains (Hall 2005).

I remember watching a beautiful short animated film, *The Man Who Planted Trees*, in the late 1980s. The film told the story of a lone shepherd, living high in the French Alps, who single-handedly planted acorns and chestnuts in an effort to restore the forest. The film was narrated by a young traveler who happened to run into the shepherd while hiking in the mountains. The traveler could not understand why the shepherd lived in solitude so high in the Alps. But after staying with him for a few days and observing the shepherd silently going about his work of planting seed after seed, the traveler came to appreciate and even love the beauty and grace in the shepherd's activities. The traveler left but returned several times over the years, right after World War I and again after World War II to see first a hillside covered with saplings and later a mountain forest. I didn't realize at the time that the film was based on a short fictional story by Jean Giono, published in 1953. I also didn't know that the story so neatly traced the history of restoration in the Alps, a history I was completely unaware of. The film is certainly more poetic than reality, as a single person – rather than a government agency – takes on the duty of repairing the land and replanting the forest, but the story was so simple and powerful that ever since it has served as a source of inspiration as I do my own work in restoration.

In the late nineteenth century in the American West there were also efforts to restore mountain forests and meadows due to an increase in flooding and erosion and a decline in rangeland for cattle and sheep, although in America restorationists focused early on the idea that human misuse of natural resources was leading to environmental decay. In western states like Utah, restorationists attempted to set up conditions that would allow nature to repair itself. Humans were seen as knocking the environment off course, and if humans added in the correct factors – planted the right species, erected structures to reduce erosion – then nature would be able to correct the damage caused by humans. Only later, in the early twentieth century, did American restorationists in the mountainous west start to think that restoration was an activity that would require constant human work in order to be successful (Hall 2005).

Environmentalism is often thought to have originated with the rise of city-states 3,000–4,000 years ago. Once towns grew into cities the problems inherent with large numbers of people living together became abundant and obvious – especially problems related to sanitation brought about by the accumulation of sewage and garbage. Thus, early in the history of cities there were complaints about water pollution and calls for ways to correct those problems, which resulted in the construction of the first sewage systems and garbage dumps (Hunter and Gibbs 2007). The use of fossil fuels, especially coal, and the Industrial Revolution led to declines in air quality so that by the 1600s there were complaints about air pollution, a belief that air pollutants led to poor human health and citizens asking for an end to air pollution (Hall 2005). Although these early signs of sensitivity to environmental degradation did not relate directly to ecological restoration, in many cases the causes of environmental problems in cities led to the degradation of nearby ecosystems that were so severe that the citizens began to demand restoration of the damaged ecosystems. Sometimes the attempt to correct one problem (such as waste disposal to improve the sanitation of cities) led to environmental problems elsewhere. One of the greatest sources of environmental degradation was the flushing of large amounts of sewage into rivers and wetlands near cities. When a small amount of sewage is flushed into an aquatic ecosystem, the ecosystem is able to absorb it and decompose it. However, large amounts of sewage, often accompanied by garbage being tossed into the river or wetland, overwhelm the ability of the ecosystem to absorb and decompose it. Thus rivers and wetlands in and near cities often became fetid, stinking pestilent places with foul-smelling air and water so putrid that drinking it would make a person ill.

Citizens and local governments began to demand changes to sewage treatment that would restore rivers and wetlands to their previous, more healthy and better functioning condition. In the middle of the nineteenth century massive projects were undertaken in London to treat sewage before it flowed into the River Thames and its tributaries in order to restore the river. Clean-up of the Thames and its tributaries was due directly to the Great Stink of 1858 and earlier outbreaks of cholera in the city. As a result, the construction of sewers and embankments commenced in 1859 to quickly remove waste and increase the rate of flow of the rivers in order to quickly flush filth from the river system. However, problems with water pollution were evident long before the Big Stink, and proposals to construct sewers in London had been made at least as early as 1834 (Halliday 1999). In 1878 Frederick Law Olmstead was hired by the city of Boston to help design restoration and management plans for the Back Bay Fens in order to eliminate or at least reduce water pollution and environmental degradation due to sewage flows in the Fens. Although the plan Olmstead developed did not call for restoration to the exact original wetland that once existed in the Fens, he developed a plan that would restore natural ecosystem function and create a salt marsh and surrounding vegetation that was similar in species composition and hydrology to the original. Olmstead specifically thought of his plan as a restoration project; the restoration of the Back Bay Fens is a candidate for the first large-scale restoration project in North America (Egan 1990). Thinking about reducing water pollution due to sewage inputs as restoration may be a nineteenth-century innovation, but the project of trying to alleviate the stresses to aquatic ecosystems caused by excess sewage is an activity that may be almost as old as cities themselves.

The beginnings of restoration science

Academic study of ecological restoration almost certainly began in the late eighteenth and early nineteenth centuries as European foresters and hydrologists designed experimental plantings to improve forest health and to reduce erosion and limit damaging run-off and flooding. Their efforts at reforestation were experimental in the sense that no one knew how well they would work and that the restoration projects were observed and measured for many years. They were not experimental in the sense of a replicated design with control treatments. But even given the limitations in experimental design that are apparent to experimental ecologists today, the fact that theories about how to repair damaged ecosystems were being developed and tested was a huge step forward (Whited 1996; Hall 1997,

2005; Dudley *et al.* 2005; Fischer and Fischer 2006). A major limitation to these early efforts was that many foresters relied on planting trees that were easy to obtain and propagate, with little concern about whether the trees were native or not and with no attempts to restore other parts of the former forest biota (Fischer and Fischer 2006).

A similar experimental approach to ecological restoration took root in the American West, especially in the mountains and rangelands of Utah and eastern Oregon around the turn of the twentieth century. Many of those restoration studies were carried out by foresters, range managers and hydrologists trained at land grant universities, some of whom, such as Arthur Sampson, returned to teach at those land grant universities after spending many years working in the field (Hall 1997).

The link between experimental studies of restoration in Europe and the United States was George Perkins Marsh. Unfortunately, Marsh is in many ways the great forgotten figure of environmental studies and restoration ecology. Many restorationists who speak of Aldo Leopold with awe and reverence have almost no knowledge of Marsh's work or his role in bridging the traditions of restoration as practiced in Europe and the United States. Marsh was an exceptionally gifted linguist, fluent in 12 languages, who was widely read and traveled (Hall 2005). He grew up in Vermont and observed degradation to the landscape there and throughout New England. He correctly deduced that erosion and loss of soil fertility were due to excess logging, overgrazing and poor farming practices. He thought that if humans damaged the landscape, humans were also capable of repairing it. In a seminal speech in 1856 at the New Hampshire State Fair he explained how humans had harmed New England's lands and, more importantly, explained ways that humans could work to repair the land. By that time he had already traveled extensively throughout Europe, especially the Mediterranean region. In 1861 he returned to Italy to serve as the US ambassador there, a position he would hold until his death in 1882. Early in his tenure as ambassador, he finished and published his landmark book, *Man and Nature* (1864), often considered to be the first book focused on environmental studies. In *Man and Nature* he provided numerous examples of human degradation of the landscape and also described ways humans were working to restore and improve the damaged land (Hall 2005).

Marsh was a synthesizer who spent much time traveling and talking with people as he learned more and more about environmental degradation and ways to possibly correct these problems. These observations and conversations influenced him as he developed his vision of how humans might restore the landscape. Marsh understood that the declines he and others observed were caused by poor human land use and he thought that with proper human management and modifications to the landscape, natural landscapes would be able to heal themselves. Marsh's most important insight was that human efforts to improve the land, such as by introducing new species or straightening streams to improve drainage, often have unintended consequences that actually result in degradation to the land. While that idea seems like common sense to us, to his contemporaries it was a revolutionary notion. Marsh continually emphasized both the human responsibility and ability to fix those problems if we can correctly interpret the landscape to see how it functioned in the past and will best function in the future (Hall 2005).

Marsh achieved a trans-Atlantic cross-pollination of ideas because *Man and Nature* was widely read on both sides of the Atlantic and highly influential in both areas. European officials began to realize that human actions were causing many of their environmental problems, rather than some kind of inevitable decline in an unmanaged nature. At the same time, Marsh's book helped to change Americans' perceptions of their impacts on the land because there was increasing recognition that the American landscape, especially the arid West, could not support typical American land-use practices. As early as 1878 John Wesley Powell wrote that the West could not support the level or type of agriculture and development used in the East. However, his call for caution in the development of the intermountain West was largely unheeded and ranchers and loggers moved in to exploit the resources of the area. By 1888 it was apparent that overgrazing and land clearing were leading to rapid degradation of forest and mountain ecosystems in the West, with Utah singled out as suffering especially great amounts of damage (Hall 2005).

In 1902 Alfred Potter, the US Forest Service's main administrator in charge of rangelands, reported that soil erosion was severe in Utah and that restoring the vegetation was essential to saving grazing lands there. His observations led to forest rangers beginning reforestation in Utah in 1907 and to the creation of the Great Basin Experiment Station in 1912. The Experiment Station was dedicated to finding ways to restore the intermountain West – especially to restore watersheds to their former status as productive grasslands and forests and to help prevent continued flooding and soil erosion. The Experiment Station and its scientists, together with other foresters and range managers based throughout the West, especially in Oregon's Wallowa Mountains, New Mexico's Jornada Range Reserve and Arizona's Santa Rita Range Reserve, conducted extensive research into forest and rangeland restoration over the next 50 years, and these sites continue to be important research areas today. At first, many of these scientists and managers attempted to restore ecosystems using easily obtained and fast-growing non-native species. They wanted to create especially productive ecosystems that would be most beneficial to humans. But many of the non-native species did poorly in the arid West. During the 1920s and 1930s they came to realize that the best species to use in restoration of their sites were native species. The native species provided better habitats for wildlife and withstood environmental variability better than did the non-native species. The range manager's goal became the recreation of the original ecosystem rather than the creation of a new, human-designed ecosystem. One of the early directors of the Great Basin Experiment Station, Arthur Sampson, left in 1922 to become a professor of range management at the University of California, Berkeley, where he taught generations of future range managers and emphasized the importance of restoration using native species (Hall 2005).

Restoration as an academic/scientific enterprise

So, the received wisdom when I began working in restoration – that ecological restoration originated at the University of Wisconsin Arboretum in the 1930s – was incomplete. Clearly, ecological restoration was an activity that began long before those pioneer tallgrass prairie restorationists started scattering seeds across the earth at the Arboretum. Even if we make the claim, as is sometimes done, that ecological restoration as an academic and scientific enterprise began at the Arboretum, we should recognize that such a claim also fails to acknowledge the true history of restoration. Arthur Sampson began teaching students about range and forest restoration at Berkeley in the 1920s. The Dutchess County Botanical Garden at Vassar College was another site of early restoration work. Edith Roberts and her student Margaret Shaw conducted a survey of Dutchess County's native vegetation in 1921–1922 and initiated planting native species at the botanical garden in 1924. Their goal was to re-establish the plant communities that existed at the time of Euro-American settlement, with the native plants growing in the proper environmental conditions for each community on a four acre plot at the Vassar College campus (Egan 1990). Their work neatly presaged the goals of the planners of the University of Wisconsin Arboretum, who about ten years later set about re-establishing the pre-settlement plant communities native to Dane County, Wisconsin, all growing in the proper environmental conditions on the much larger 1,260 acres of the Arboretum.

Why, then, if it is so obvious that ecological restoration did not begin in Madison in the 1930s, does the myth of restoration springing from the minds and hands of Leopold, Fassett and Sperry persist to this day? There are several reasons. The presence of Aldo Leopold as one of the originators of ecological restoration lends a certain staying power to the story. The science of ecology and the environmental movement do not have that many heroes, much less heroes that are shared by both, but Leopold is such a hero. George Perkins Marsh, Arthur Sampson and Edith Roberts are largely unknown outside of academic circles, but due to the continuing influence of Leopold's masterwork, *A Sand County Almanac*, he remains very well known and is regarded as one of the major founders of conservation-environmental thinking in North America.

Aldo Leopold developed a completely new way of examining the ethical responsibilities of humans toward the land. J. Baird Callicott (1990) described Leopold's contribution as an Evolutionary-Ecological Land Ethic. The most important aspect of Leopold's thinking about a land ethic, and what makes him such an important hero for so many ecological scientists, is that he explicitly developed his ideas from a background in evolutionary biology. He saw humans as just one member of the vast web of living things and claimed we have a responsibility to recognize the rights and values of all other living things. He began his radical, visionary essay "The Land Ethic" by recounting the story of Odysseus returning home from his wandering, and hanging all the slave girls he suspected of improprieties while he was gone. For Odysseus and his fellow Greeks, hanging the slave girls was no problem because he was simply disposing himself of some bothersome property, rather like throwing out the trash. As Leopold noted, today we think it is wrong to view people as property who can simply be disposed of, and he argued that we should stop thinking of the land as simply property to be treated as we see fit. His use of evolutionary biology and ethical reasoning to claim we have a responsibility to treat all of living creation with respect and as an equal is a huge shift in human thinking and is key to ecological restoration. His land ethic gave impetus to restorationists and inspired them to think of themselves as acting in a role as members of the ecological community, dedicated to nurturing and safe-guarding it.

The story of ecological restoration beginning at the University of Wisconsin also has appeal to native Midwesterners like me because, in this case, we see ourselves being led by other Midwesterners in a bold new enterprise that started here and spread around the world, an enterprise we think is crucial to developing better relationships between people and the environment all across the globe. The story of a Midwestern birth to a new, worldwide movement may not get much traction outside of the Midwest, but it is a powerful source of pride to us here. In many ways, it was Midwesterners, inspired by the early attempts at prairie restoration, who began to spread the word, sometimes as fervently as if it was the gospel, to other parts of North America and the rest of the world, claiming ecological restoration is necessary and is the way of the future. They spread the word that the future ecological health and integrity of the world depended not just on preserving pieces of wild nature, but on the practice of restoring and re-wilding damaged and degraded ecosystems. They felt that preservation by itself would not save enough wild land to make a difference in the long-term degradation of a planet increasingly dominated by human activities (Vitousek *et al.* 1997). Ecological restoration thus was the key to both repairing ecosystems and to developing a new, better relationship between increasingly urbanized people and the rest of the environment (Jordan 2003).

But the origination of ecological restoration with a hero like Leopold and local pride by almost evangelical Midwestern restorationists is not enough to explain the persistence of the story that restoration began in

Madison, Wisconsin in the 1930s. The story persists because although it is an incomplete story, it does contain a kernel of truth. The key element of truth in the story is that although ecological restoration has scientific and academic roots that extend much further back in time and further afield than Wisconsin, what happened in Wisconsin was both a coming together of several ideas about restoration and then an amplification of those ideas due to the development of a community of dedicated restorationists working in the area. Beginning with the planting of the Curtis Prairie in 1936, we can follow the history of ecological restoration through a series of phases. The different phases of ecological restoration are not completely separated by time or even distance, but each phase evolved in such a way that it influenced subsequent phases. What began at the Curtis Prairie was really the beginning of the first phase of modern ecological restoration.

Phase I, Track I: prairie restorationists spread the word

The first phase of restoration can be divided into two parallel tracks, with the first track blooming in Wisconsin due to the presence of prairie restorationists. The first prairie restorationists did everything by trial and error, planting locally collected seed sometimes, transplanting individual plants in some cases and even transplanting entire blocks of prairie sod in some areas. Because no one really understood prairie ecology at the time, the trial-and-error method sometimes led to mistakes. At the beginning the prairie restorationists did not use fire to maintain and rejuvenate their prairies because early twentieth-century American ecologists thought fire in general was a destructive force. But as the restored prairies soon became overrun by cool-season grasses like Kentucky blue grass, they experimented with fire and realized that fire was vital to the ecology of prairies (Curtis and Partch 1948).

Prairie restoration was practiced almost exclusively at the Arboretum for about 20 years. The first recorded example of prairie restoration away from the Arboretum was initiated at the Green Oaks Field Research Center, owned by Knox College in western Illinois, in the spring of 1955 (Figures 2.1–2.2). Two Knox College professors, Paul Shephard (later to become well known as an environmental philosopher) and George Ward visited the Arboretum in the fall of 1954 and realized they could also restore prairie at Green Oaks. They were given a supply of prairie seed by the Arboretum staff and they used that seed and seed collected locally in western Illinois in their initial seven-acre planting of East Prairie. They later expanded and began planting an area known as West Prairie. Paul Shephard left Knox College in 1963 and was replaced by Pete Schramm in 1965. Schramm immediately recognized the value of what Shephard and Ward had started and continued their work, greatly expanding their reconstruction work in West Prairie and adding in another site, South

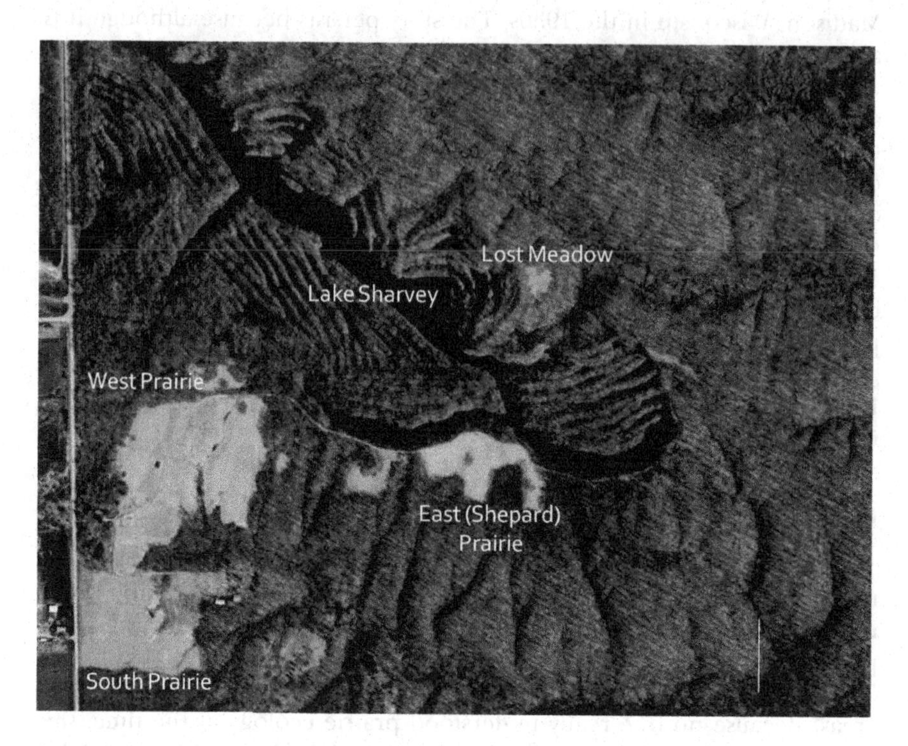

Figure 2.1 An aerial photograph of the Green Oaks Field Research Center.

Note
Green Oaks is 704 acres in area and most of it is forested with second-growth oak-hickory woodlands. East, West and South Prairies are all prairie restorations. Lost Meadow is a prairie remnant. The ridged land around the lake was surface mined for coal in 1940 and 1941. The lake lies in the final, unfilled cut of the coal mine.

Prairie. Eventually the prairie reconstructions at Green Oaks grew to be 40 acres in size (Schramm 1992; Allison 2002).

Schramm got in touch with a network of other prairie enthusiasts located in Illinois and soon began to work and exchange ideas with other early restorationists, including Ray Schulenberg at the Morton Arboretum and Robert Betz at Northeastern Illinois University. Schulenberg began the 100-acre prairie restoration at the Morton Arboretum in 1962. Betz began doing prairie restoration in the center of the Fermilab site at Argonne National Laboratory in 1971. That restoration eventually grew to be over 700 acres in size, and until the mid-1990s was the largest prairie restoration in existence. Schramm's interactions and friendship with Betz, Schulenberg and others led to the first North American Prairie Conference, hosted at Green Oaks in 1968. That was the first meeting exclusively devoted to consideration of prairie ecology and restoration and led to a long-running, successful set of biennial meetings. Many restorationists got

Figure 2.2 The founders of Green Oaks, early spring 1955.

Note
From left to right: Henry Green, George Ward, Alvah Green and Paul Shepard. Alvah Green owned the property that was acquired by Knox College in 1957. Henry Green was his brother. Ward and Shepard were professors at Knox College who started the prairie restoration at Green Oaks in the spring of 1955.

their first exposure to the practice and techniques of ecological restoration at the Prairie Conferences.

William Jordan III was hired by the University of Wisconsin Arboretum in 1977 and asked to direct their public outreach and publications. Jordan didn't know anything about ecological restoration when he arrived there, but he quickly realized that ecological restoration provided a new way of both repairing damage to the landscape and a new relationship between humans and the environment. He spent long hours exploring the restorations at the Arboretum and talking with staff there, especially Keith Wendt and Brock Woods. Jordan was inspired by what he saw at the Arboretum and what he heard was happening at places like the Schulenberg Prairie and Fermilab. He had his ear to the ground and soon learned that prairie restoration was happening in dozens of small sites across the Midwest. He felt there was a need to provide an outlet for information about restoration and a place for restorationists to exchange ideas, so in 1983 he convinced the University of Wisconsin to begin publishing *Restoration and Management Notes*, the first journal in the field (it has since been renamed *Ecological Restoration*) (Jordan 2003). *Restoration and Management Notes* was in many ways an idiosyncratic publication that reflected the interests and tastes of Jordan. It contained basic how-to articles about restorations, news of newly established restorations and also became a forum for people to exchange ideas about the ethical, political and social dimensions of ecological restoration. Every issue, until Jordan resigned as editor in 2002, contained an editorial by Jordan that almost always included encouragement and exhortations to continue the good work of ecological restoration.

Phase I, Track II: mine reclamation to restoration

The second track in the first phase of ecological restoration developed at about the same time as the first track. But there is a key difference between the two tracks. Track I, exemplified by the prairie restoration enthusiasts, was primarily concerned with recreating previously existing, historical ecosystems that had been damaged, lost or destroyed, often working in abandoned agricultural fields or land that was set aside because it was of marginal value for development. But in Track II, best exemplified by people working in mine reclamation, the main emphasis was on taking land that was utterly destroyed, often without any ecological value, and restoring that ecosystem so that it had improved ecological function. In the second track, restoration to a previously existing historical ecosystem was less important than simply reducing the damage to a particular site and creating an ecosystem that provided some functions and services of value to either humans or the greater environment or both.

It should not be surprising that many of the early practitioners of mine reclamation were located in Europe and that many of the government regulations requiring restoration of damaged mines or other industrial sites were first enacted in Europe. Mining practices for many different materials – peat, coal, iron and other metals, clay and gravel – have occurred in Europe for several centuries. The Industrial Revolution began in Europe and industrial production and the resulting environmental degradation developed earlier in Europe than in the rest of the world. Given the high population densities in many European countries and the relatively small geographic area of most European countries (especially when compared with countries like the United States, Canada and Australia), it was inevitable that degraded damaged land, often termed derelict land (Bradshaw and Chadwick 1980), would be regarded as not just an eyesore or even as unhealthy, degraded land, but as a waste of space, space that was needed for other activities. Thus restoration of that land to alleviate damage, remove unhealthy environments and create newly useful land was seen as highly desirable at about the same time that restorationists were beginning prairie restoration efforts at the Arboretum in the 1930s and 1940s. In the United Kingdom the Town and Country Planning Act of 1944 gave local authorities the right to acquire damaged or derelict land and restore it to a useful condition. That act was revised and expanded by Section 89 of the National Parks and Access to the Countryside Act of 1949 (and revised again in Section 6 of the Local Authorities Land Act of 1963). Several similar laws were also passed in the United Kingdom during the mid-twentieth century. In 1951 the Mineral Workings Act set up guidelines for creating a fund that would provide financing for the purchase and restoration of land damaged by open-cast ironstone mining. The Mine and Quarries (Tips) Act of 1969 required local authorities to prevent abandoned tips (tailings sites) from being dangerous to the general public due to instability. Restoration of the abandoned tip was one way of ensuring that the tip was no longer dangerous to the public (Bradshaw and Chadwick 1980).

The United Kingdom was not the only country to enact legislation to allow or even require land restoration. In Germany, a law passed in 1920 allowed for the creation of a regional planning association in the Ruhr region. In 1962 that law was expanded to create a more powerful regional planning group that was given the responsibility for developing land-use plans and for determining how to solve problems caused by degraded lands and refuse disposal (Bradshaw and Chadwick 1980).

The interesting point about the context of land reclamation in Europe is that the focus was clearly on taking land that was determined to have no ecological or biological value due to a loss of topsoil, the deposition of huge piles of mine tailings or industrial slag, high rates of erosion and toxic soils poisoned by heavy metals (with subsequent toxic heavy-metal-polluted run-off) and then to restore that land to a functioning ecosystem that provided some kind of value to society and the environment. The mindset of restorationists such as Anthony Bradshaw seems to have been that any kind of functioning ecosystem was better than derelict land. Restoration or reclamation work was done by engineers, hydrologists and landscape architects, who started with the goal of producing whatever kind of ecosystem was possible on that derelict land. If the land supported grassland before being degraded but due to excavation the surface was below the water table, then creating a marsh or pond was an acceptable solution. Or perhaps a former forest site that was damaged might be restored to support agriculture. The development of recreational sites such as urban parks or golf courses might be the best restoration goal in a more urbanized or developed area (Bradshaw and Chadwick 1980).

In North America the reclamation phase proceeded by fits and starts. The first serious reclamation efforts in the mid-twentieth century in the United States were directed at ending the soil erosion that created the Dust Bowl years of the 1930s. The Soil Erosion Service (SES) was created as part of the Department of the Interior in 1933. It was transferred to the

Department of Agriculture in 1935 and soon after that renamed the Soil Conservation Service (it was renamed yet again in 1994 when it became known as the Natural Resources Conservation Service). The original mandate of the SES was to develop ways to prevent soil erosion, especially soil erosion due to poor farming practices, and then to educate farmers and ranchers so that they would farm in ways that limited soil erosion. The agency had a great deal of success in achieving those goals (Hunter and Gibbs 2007; www.nrcs.usda.gov).

Mine reclamation developed somewhat later in the United States. This may be partly due to the large size of the United States and the fact that mine operations were often located either far from urban areas or in areas with few – and frequently poor – people. Thus, many people in the United States were either not aware of the damage caused by mining or felt the country had enough good land that it could afford to have some areas that were damaged in the production of goods and the extraction of resources. The Federal Surface Mining Control and Reclamation Act of 1977 was enacted to regulate surface mining for coal (and only coal). The Act required that lands surface mined for coal after 1976 had to be restored to the original contour of the land. During mining the topsoil was to be removed, stored and then placed on the site again. Vegetation similar to what existed prior to mining was to be planted during restoration, and damage to local hydrology had to be limited (Bradshaw and Chadwick 1980; Hunter and Gibbs 2007). Other types of surface mining in the United States do not have the same kind of reclamation requirements and there are loopholes that allow mine operators to avoid restoring to the original contour for coal mines. However, there has been much restoration of surface-mined lands since 1977 in the United States.

One of the most spectacular examples of land degradation due to environmental pollution is located near Sudbury, Ontario in Canada. The richest deposits of nickel ore in the world exist in the Sudbury region and industrial mining of the copper-nickel ore began there in 1886. The combination of slag deposition and fallout from the nickel smelters created a huge area of environmental devastation. Forty square miles suffered complete elimination of all plant life, about 140 square miles supported only scrubby shrubs and herbaceous plants, and about 1,700 square miles suffered some kind of damage to plant life due to acid rain and metal toxicity (DeLestard 1967). Attempts to restore the damaged landscape began as early as 1947 in Copper Cliff and additional efforts were made in the 1950s, but these all met with very little success. It was not until the 1960s that any real progress was made in restoring the damaged lands; restoration efforts continue to this day (Winterhalder 1984).

These Track II reclamation efforts in many ways fit in with the long tradition of attempts to prevent erosion and restore forest and grazing lands that began in France, Italy and the western United States in the nineteenth century. But these efforts at land restoration differ in the type of land being restored and the approach to the problem. The reclamation workers were concerned with restoring land that frequently had been utterly destroyed – topsoil removed and lost, contour of the land drastically changed and many toxic substances released to the soil, water and air. They also approached the problem of restoration as primarily an engineering problem that could be solved by replacing damaged soils, re-establishing a more natural topography and hydrology all with the goal of restoring at least some ecological functions and value to the land. The Track I restorationists were mostly ecologists, field biologists, wildlife biologists and botanists interested in restoring the land to a historical, pre-existing condition. Track II reclamation scientists were more focused on the idea that almost any ecosystem would be better than the derelict, damaged sites they were asked to reclaim or restore.

The two groups at first had little contact or communication, but that changed in the late 1970s and 1980s. Many scientists who were first exposed to restoration through mine and industrial site reclamation became leaders in the field of ecological restoration as it developed in the 1980s. Anthony Bradshaw from the United Kingdom, John Cairns from the United States and Keith Winterhalder from Canada were especially important in introducing the practice, successes and challenges of reclamation to the ecological restorationists through their many books, papers and oral presentations. In 1980 Bradshaw and Cairns independently published books (*The Restoration of the Land: The Ecology and Reclamation of Derelict and Degraded Land* co-authored by Bradshaw and Chadwick, and *The Recovery Process in Damaged Ecosystems* by Cairns) that became landmarks in the field as they provided both a summary of the state of the art of ecological restoration at the time and also proved to be inspirational to many scientists and lay practitioners who were just becoming involved with restoration projects.

The first phase of ecological restoration drew to a close with the founding of the SER in Madison, Wisconsin in 1988. The SER was founded with the idea that it would serve as an intellectual home for both academics working in restoration ecology and practitioners working in the field. The society has become a dynamic organization that promotes interactions by member restorationists from all over the world. Although the membership is on the small side – about 2,000 members in 2010 – the members are active and committed to promoting good ecological restoration and providing outreach to a global public about the benefits of ecological restoration. We will come back to the work of the SER throughout this book.

Phase II: the birth of citizen-led restoration

The second phase of ecological restoration was the expansion of the field beyond engineers, landscape architects and academic ecologists to the general public, as members of the lay public became increasingly active as volunteers at established restorations (such as at the Morton Arboretum) and even began to initiate their own restoration projects. Ironically, in terms of my own personal history, one of the initial events in this expansion of restoration to a more public, less academic activity, happened in the Midwest just as I left to go to college. In the fall of 1977 I headed off to the University of Puget Sound, with plans of becoming a marine biologist. A bit earlier in 1977, Steve Packard began working in the Chicago area on what would eventually become known as the North Branch Prairie Restoration Project. Largely due to Packard's efforts, the Chicago area became a hotbed of restoration activity where, by the mid-1990s, on any given weekend as many as 30,000 people would be working on projects aimed at restoration of prairie, savanna, wetlands and forests (Gobster and Hull 2000). But I was unaware of any of that activity until I returned to the Midwest in the fall of 1993, when I saw how large a movement ecological restoration had become. I came back to the Midwest to teach at Central College in Pella, Iowa and soon after my return I was introduced to Pauline Drobney at the Neil Smith National Wildlife Refuge (NWR), where a very large prairie restoration project was underway (the eventual goal is to restore 8,000 acres of former agricultural land to tallgrass prairie). I did some volunteer restoration work with Pauline and followed her around Neil Smith NWR, simply amazed by the spectacular scale of the project. I was also amazed by how many volunteers came out to work at Neil Smith and at how many of the volunteers were also involved with smaller prairie restoration projects – often projects the volunteers were conducting on their own land.

When I was growing up in the Midwest, prairie plants, when they were considered at all, were considered to be unattractive weeds that took up valuable land that might be better used for farming (Figure 2.3). I remember my grandfather following the advice of the then Secretary of Agriculture Earl Butz (serving from 1971 until 1976) that farmers should plant from fencerow to fencerow, eliminating any competing fringes of native or non-agricultural plants. Thus he had the thick hedgerows of osage-orange pulled up on the family farm. Before their removal I remember walking along those hedgerows in the fall, kicking the large "hedge apples" shed by the trees and marveling at the size and density of those hedges. Removal of the hedges was a decision my grandfather realized too late was a mistake due to increased wind erosion and the loss of many small animals (quail, rabbits, songbirds) from the countryside. So I was surprised, even shocked, to find so many Midwesterners, many of them farmers, proudly cultivating prairie plants on their land.

Pauline introduced me to Daryl Smith from the University of Northern Iowa. In Daryl I met a sage prophet of prairie ecology and restoration who had convinced the Iowa Department of Transportation to plant many miles of roadside with native prairie vegetation. Clearly there had been huge changes in the perceptions of many Midwesterners during the 16

Figure 2.3 Purple prairie clover (*Dalea purpurea*), a once unloved native prairie plant that is now considered both highly desirable and attractive.

years I was off pursing higher education and additional scientific training. Once I became aware of the breadth of ecological restoration I started reading the journal *Restoration and Management Notes*, and many articles published there made it obvious that ecological restoration had become a public activity across all of North America, in many parts of Europe and Australia and even in some developing nations. The 1980s and 1990s were a time of rapid growth in the practice and acceptability of ecological restoration.

Two other key events occurred during this second phase of ecological restoration. First was the increasing contact and communication among academic restoration ecologists, professional restorationists (often engineers and landscape architects) and lay practitioners of ecological restoration. The contact and communication was facilitated by the SER and journals like *Restoration and Management Notes*, which was explicitly edited by William Jordan to enhance dialogue among those groups, and *Restoration Ecology*, which was more focused on promoting a rigorous approach to the scientific study of ecological restoration. Communication among the groups was important in the exchange of ideas, techniques and even inspiration so that all three groups became more committed to the practice of ecological restoration. However, from the beginning of Jordan's efforts and the formation of the SER to the present, some people have expressed concern that practitioners and academics do not communicate well at all (Cabin 2011). I will return to that issue in the final chapter.

The second key development was a refining of what ecological restoration means and the goals for restoration projects. The early academic restoration ecologists, certainly those at the University of Wisconsin Arboretum and Vassar College, started with the idea of using restoration to recreate the ecosystems that existed at those sites at the time of Euro-American settlement. But this focus on historicity raised two issues: wouldn't those ecosystems have changed even if Euro-Americans hadn't arrived (given our newly developed understanding that ecosystems are dynamic, always changing (Botkin 1990))?; and how could we apply standards of historical restoration to ecosystems on continents where there is not an obvious arrival of new people (such as Europeans coming to America or Australia) and subsequent change in land management? This led to a further question of whether the focus on Euro-American settlement as creating a change from the natural order did not in fact lead to a denial of the vital role played by native peoples in managing the landscape?

The reclamation professionals originally focused on a more pragmatic approach of trying to bring any kind of positive function to a damaged ecosystem, but there was a shift by them to a greater attempt to restore an ecosystem to be as similar to the original as possible. But the critical shift for both groups was a recognition that the most important goal for ecological restoration should be to restore the ecosystem to the point at which it could function as a dynamic, changing ecosystem where ecological and evolutionary processes could occur with as little hindrance from human activity as possible. The shift to thinking about ecological restoration as the restoration of a process more than a particular product is reflected in the 2002 SER definition of restoration as an activity that restores health, integrity and sustainability to an ecosystem. Although some restorationists still focus on attempting to recreate a particular historical ecosystem, throughout the 2000s that approach became less common and the focus on process within a local context became more accepted.

Phase III: restoration, climate change and the rise of novel ecosystems

Today we are entering a third, new phase of ecological restoration. This third phase of ecological restoration has arisen as a response to concerns about the effects of global climate change on both restored ecosystems and the earth as a whole. Restorationists are also concerned that human habitat modifications and the spread of non-native species are leading to the world becoming increasingly domesticated (Kareiva *et al.* 2007) and homogenized (McKinney and Lockwood 1999). Non-domesticated ecosystems continue to suffer greater and greater amounts of habitat fragmentation (Fischer and Lindenmayer 2007).

At the turn of the millennium Richard Hobbs and Jim Harris claimed that ecological restoration will be the most important tool for preserving and maintaining species and ecosystem diversity in the twenty-first century. They issued a call to all restorationists to use ecological restoration to help solve the many ecological problems facing the earth. And further, they recognized that if ecological restoration is to fully realize its potential in the twenty-first century, restorationists will have to be adept at addressing the benefits that can be achieved via restoration both to the ecosystem and to the human stakeholders involved, and this will mean integrating both ecological and cultural considerations into restoration planning and implementation (Hobbs and Harris 2001). Whether restorationists will be able to achieve their goals in the twenty-first century will be almost entirely dependent on how restoration projects and indeed the entire planet are affected by the twin specters of increasingly rapid rates of environmental change (driven primarily by global climate change) and the development of novel ecosystems. Hunter Lovins has referred to these rapid changes as "global weirding" because many of the potential changes to global climate and ecosystems are unpredictable given our current understanding of global ecology, and because the novel ecosystems will result in combinations of species which are new to us in ways that seem, for lack of a better term, weird (Waldman 2008).

Novel ecosystems can be defined as ecosystems dominated by new combinations of species (usually native and non-native to the region) that have not previously existed. They probably originated as a result of some degree of human introduction of non-native species, whether purposeful or accidental, but the non-native species became so well established that novel ecosystems are now self-perpetuating and do not require human agency for their maintenance (Hobbs *et al.* 2006). Novel ecosystems have the potential to change local ecosystem structure and function, sometimes in desirable ways but sometimes in undesirable ways. At the 2009 meeting of the SER in Perth, Australia there were many papers presented which discussed novel ecosystems and they were a constant topic of conversation in the hallways in between talks. They were a hot topic at the meeting because of our realization that such ecosystems are becoming more common and are almost ubiquitous, and because recent experience has taught many restorationists that it will be almost impossible to completely eliminate non-native species from ecological restorations. Therefore we will almost certainly be producing restorations that result in some form of novel ecosystem, but our goal will be to end up with novel ecosystems that help preserve both biodiversity and desirable ecosystem structure and function.

Despite just recently becoming a hot topic, novel ecosystems are not new. My father's parents retired to Santa Barbara, California before I was born. My family and I made almost annual visits to them throughout my childhood, and as an adult I continued to visit them until they passed away when I was in my early thirties. Rather like Proust, many of my earliest memories of Santa Barbara are olfactory – the salty smell of wind coming off the ocean, the tar smell on the beaches resulting from petroleum seeps and the spicy odor of eucalyptus. When I was in graduate school at the University of California, Berkeley there was (and still is) a magnificent grove of blue gum (*Eucalyptus globulus*), outside of the Life Sciences Building. It was rumored that they were some of the tallest blue gums in the world, growing so tall because they lacked any herbivores and diseases. Eucalyptus trees grow beside roads all along California's coastal highways and in the inland valleys of the coast ranges. In fact, I cannot imagine California without eucalyptus trees. But, of course, eucalyptus are not native to California – the first eucalyptus was introduced to California by Australian miners coming to the Gold Rush in the 1850s. Today eucalyptus are so well established that many people think they are native to California, and even if they know the trees are not native, they don't want to see them removed because to them, like me, eucalyptus are a central part of their experience of California. Some people also fear that removal of eucalyptus will lead to erosion on steep hillsides and a decline in biodiversity. In contrast, native plant enthusiasts in California would love to see eucalyptus permanently removed from the state, regardless of how difficult that would be to accomplish (Coates 2007). The dominance of eucalyptus in California is hardly unusual for a novel ecosystem, but it stands out because the trees came from Australia and because they are so striking in appearance and aroma that they can't be missed. But back home in Illinois, I live in an ecosystem in which the most common grass is Kentucky blue grass (*Poa pratensis*), native to Eurasia, and two of the most common birds are house sparrows and European starlings, both native to Europe. We can find similar examples on virtually every continent and in most nations.

At the 2009 SER meeting in Perth I had a long talk with Richard Hobbs of the University of Western Australia about novel ecosystems. He sees them as ecologically interesting and exciting – he thinks there is much we need to learn about them and, if managed correctly, they have the potential to help us preserve biodiversity and ecosystems that would otherwise be lost. He is also a pragmatist who thinks that we will have to work with novel ecosystems because, given the current trajectories of species introductions and global climate change, novel ecosystems are the only ecosystems we may have in the not-so-distant future. I'm less sanguine about them and worry that novel ecosystems will lead to a homogenized world in which the same species – starlings, Kentucky blue grass, eucalyptus, etc. – are found everywhere. I have heard Dan Janzen describe such a future world as ecologically boring, and he is right if the world really does become dominated by a cosmopolitan cast of a few hardy species that tolerate human disturbance regimes.

But the third phase of ecological restoration is not just about a consideration of restoration in the context of global climate change, novel ecosystems and general global weirding. The third phase is about a continued expansion of what ecological restoration means and why it is an activity worthy of increased human involvement, whether by individuals, citizen groups or governmental agencies. For many restorationists, the primary goal of restoration has been to restore damaged ecosystems to their previous, undamaged condition (Higgs 2003; Jordan 2003). For some restorationists, such restoration projects are still the main goal (Jordan 2003). However, there is an increasing focus on a more pragmatic type of restoration focused on restoring vital ecosystem services even if the restored ecosystem is not exactly or even remotely similar to what existed previously (Hobbs and Harris 2001; Seastedt *et al.* 2008).

Another vitally important and interesting change in ecological restoration is the development of restorations that more explicitly address human needs. Pandey (2002) called for restorationists to engage in "holistic restoration" in which the goals of the restoration project would be to restore native biodiversity, full ecosystem function, ecosystem services and also the link between nature and society. Clewell and Aronson (2007) later used the term holistic restoration in a slightly different way to mean restoring "wholeness" to an ecosystem – restoring the ecosystem in its entirety so that all parts and functions are present. Clewell and Aronson's use of the term "holistic restoration" acknowledges the need for restoring the link between nature and society and providing ecosystems services, and they place special emphasis on the need to include stakeholders and all interested parties in the restoration process. But their emphasis on wholeness is slightly different from Pandey's original definition. Some of these more human-centered restorations are primarily concerned with using restoration to provide opportunities for indigenous people to manage their land using traditional methods in order to maintain their unique culture and other restorations, often in a more urban context (but not always), are focused on restoration as a way to achieve environmental justice for poor people and/or minority populations who in the past frequently were not included in the restoration process (Tomblin 2009). For some restorationists, successful restoration will only be possible when ecological restoration is used to help bridge the gap between scientific- and engineering-based projects and the socio-political context in which people as a whole live. Restoration thus becomes not just a project to provide ecosystem services, but a way to also develop a sense of community and economic opportunities for society as a whole (Scott and Whitbread-Abrutat 2010).

The third phase of ecological restoration is just beginning and in it restorationists are grappling with how to work in a rapidly changing world. We are struggling to learn how to fit our traditional focus on restorations tied to local history and ecosystem structure and function into that rapid change. Many leaders in the restoration movement think that ecological restoration will have to become a project embraced by the entire world community if it is to succeed (Harris and van Diggelen 2006). There is also a recognition that ecological restoration will have to be part of a much broader program of ecological landscape design in which ecological principles are applied to better management, development and restoration of intensively used urban, suburban and agricultural lands (Lovell and Johnston 2009). It is too early to say where the third phase of ecological restoration will go, but if it is to achieve anything, greater outreach beyond the traditional academic and reclamation fields will be necessary. Ecological restoration will have to be seen as a valuable contribution to human land management in general and as an activity that people anywhere can embrace and engage in.

3 Restoration is an active choice

In this chapter I will:

- explain why ecological restoration is a necessity in the twenty-first century;
- ask why people choose to restore ecosystems, and then examine several possible reasons, including:
  - restoring lost or destroyed ecosystems;
  - maintaining biodiversity;
  - restoring severely degraded sites;
  - improving ecosystem services;
  - a moral obligation to restore ecosystems.

Why ecological restoration is necessary

Recently I was looking at aerial photographs of the American Midwest – photographs of the landscape of western Illinois and eastern Iowa – the landscape of my childhood and current home. As I looked at the photographs, everything in view was familiar to me, even if the hawk's eye viewpoint was unusual. The landscape was divided into a grid of mile-square blocks with each block divided into various land uses – large rectangular fields planted with either soybeans or maize, the occasional farmstead with a house and a few outbuildings, or sometimes a small town that had grown beyond the edges of the surveyor's original mile-square grid and created fuzzy boundaries at their margins. And what was living on the landscape? From the aerial photographs I could see that there were soybeans and maize obviously – large fields with sharp edges between them and absolutely no intervening vegetation taking up space between the fields. The farmsteads had trees and green lawns around the house and often I could pick out a vegetable garden near the house. In the towns there were many trees, shrubs and green lawns. The straight roads that formed the sharp borders to the mile-square blocks of land had very narrow vegetated margins that were hard to decipher from the aerial view but which I knew from personal experience would be made up of a mixture of grasses that are usually mown fairly closely to prevent the emergence of unwanted weeds. Sometimes, in one of the aerial photographs I would find a stream which disrupted the regular grid and meandered across the landscape with a fringe of trees and other plants forming a riparian corridor that tightly hugged the narrow path left for the stream between the fields and towns.

Indeed, what was most striking about this landscape was that it was so completely dominated by human choice and activity. Almost every feature of the landscape, especially the living green blanket of plants, was there because it was in some way useful to humans. There were fields producing valuable crops, trees to provide shade and beauty and lawns for recreation and beauty as well. But only in the narrow, meandering riparian corridors was there evidence of life that was not there due to human design. Only along those streams was there any sign that there might be living things that existed outside of human control and desire.

Of course the entire planet is not so highly constrained by human activity. I'm currently in England on a sabbatical, living in rural northern Buckinghamshire, and one of the most striking differences between Buckinghamshire and my home state of Illinois is the more lightly sculpted English landscape. As I walk along the public footpaths across the irregularly shaped pastures, there are many hedgerows, spinneys and little woods that allow for the presence of wildlife. Even though Buckinghamshire is also a landscape highly controlled and manipulated by human handiwork, the presence of that handiwork seems softer and more willing to admit that there may be other living beings with needs that don't necessarily correspond exactly with ours.

The fact that we humans have left a huge footprint across earth is not really news anymore. But sometimes, such as when I am looking at aerial photographs of my homeland, the scale of that footprint comes sharply into focus. What is news, is that almost every day we learn more about the size of the human footprint and the degree to which our influence on the earth's biota is truly global. Our first clear, modern understanding of the human impact on the planet came with Rachel Carson's 1962 publication of her immensely important book, *Silent Spring*. Although *Silent Spring* elicited fierce criticism when it was first published, Carson's claim that the widespread use of industrial pesticides and chemicals was having a profoundly negative effect on many species has been well supported. *Silent Spring* is often credited with giving birth to the modern environmental movement and leading directly to the first Earth Day on April 22, 1970.

After Carson's work it was still possible to believe that human effects on the planet's ecosystems were limited to certain areas and were reversible if we could change our behavior. Indeed, a sense of optimism, that it was possible to change paths and restore the earth to its former state was a vital part of the early environmental movement. However, for many people, the sense that some parts of the earth may have escaped human influence was largely destroyed by the publication in 1989 of Bill McKibben's book, *The End of Nature.* McKibben's book was every bit as controversial and attracted as much criticism as *Silent Spring.* I still dislike McKibben's formulation that nature exists where there is no human influence. He claimed that evidence of human impacts on a global scale, meaning no ecosystem was free from the effects of aerial pollutants and climate change, shows that we have killed nature. I find it extremely limiting to humanity to think of nature as something that exists outside of us and I agree with many environmental historians that McKibben's concept of nature (a view he shares with many other people) is a human construct that erects a dichotomy that fails to grasp the reality of both what it means for something to be wild and the complexity of human relationships to our environment and the earth (Cronon 1995; Graber 2003; Allison 2004; Ridder 2007). Even so, McKibben created a powerful story that changed many people's (including my own) thinking about the effect we are having on the planet. Post-McKibben we have found more and more evidence of the size of the human footprint and the huge impact we are having on a global scale. Indeed, post-McKibben, the size of that impact becomes ever more frightening to consider and more daunting to change.

Even a brief and rapid recitation of what we know about human impacts on global ecosystems is staggering. Humans, just our single species, consume, directly and indirectly, about 40 percent of all the carbon fixed by photosynthesis on a global scale (Vitousek *et al.* 1986). Human activities have altered all of the earth's biogeochemical pathways (Vitousek *et al.* 1997). Two-thirds of the earth's ecosystems are thought to be degraded in some way due to direct or indirect human activity, thus reducing their ability to perform vital ecosystem services (Nellerman and Corcoran 2010). It is hard to visualize what it means for us to use such a huge proportion of the products of photosynthesis or to change all biogeochemical cycles or to suffer reductions in ecosystem services, but it is easy to see, as I did when examining aerial photographs, physical human impacts on the landscape. Human activities have significantly changed about 75 percent of the earth's ice-free land surface (Ellis and Ramankutty 2008) and more than half of the earth's total land surface has been domesticated in some way or another (Sanderson *et al.* 2002). Land that might be considered wild because it is not dominated by human activity is now restricted to a few places on earth, most of which are only marginally habitable, such as deserts, high alpine areas, far northern tundra and taiga, and the deepest interiors of tropical rainforests (Sanderson *et al.* 2002). Wild and semi-wild or semi-natural ecosystems (semi-natural ecosystems such as grazing lands still contain the majority of their native species and original ecosystem functions but are influenced by human activities) are increasingly developed and exploited by humans. Since 1950, in North America the fastest-growing land-use category has been exurban development as suburbs continually spread further and further from the central city.

Exurban spread is also exacerbated by more extended development as people live out their Thoreau fantasies by constructing either primary or secondary homes in formerly wild areas in the western mountain and northern woods. Today about 25 percent of the land in the contiguous United States consists of low-density (6–25 homes per square kilometer) exurban development (Hansen *et al.* 2005). Researchers have found that even low-density development leads to changes in species behavior, declines in biodiversity and promotes the establishment of non-native species (Hansen *et al.* 2005; Huston 2005). A continent-wide examination of North America reveals that the patterns I observed looking at aerial photographs of the American Midwest are almost continuous across North America, with the landscape dominated by rectangular fields, cities, suburbs, farms in cleared forests and pastures grazed by domestic livestock (Cardille and Lambois 2010).

Increasing human population and changes in land use, especially the increasing human domination of all ecosystems, have been largely responsible for declines in biodiversity at a global scale (Millennium Ecosystem Assessment 2005; Mooney 2010; Rands *et al.* 2010). Losses of biodiversity continue to occur at ever more rapid rates and there appears to be no end in sight for these losses (Mooney 2010; Rands *et al.* 2010). In Europe there is clear evidence that the intensity of land use, measured by determining how much nitrogen-based fertilizer was used in an area (more N-based fertilizer means more intensive land use), is negatively correlated to biodiversity (i.e., more intensive land use leads to lower biodiversity) (Kleijn *et al.* 2009).

For many years ecologists debated whether there was a relationship between biodiversity levels and ecosystem functioning, but now there is a developing consensus that higher biodiversity does in fact lead to more or better ecosystem functioning. In general, higher species richness (the most common measure of biodiversity) leads to a greater production of biomass and more resource use within trophic levels. The increase in productivity and resource use in turn leads to less variability in both of those factors over time; thus increased biodiversity leads to improved, more consistent ecosystem services for both humans and the rest of the biota (Duffy 2009). There is a plateau beyond which more diversity does not lead to further increases in ecosystem function. Beyond that plateau, additional species may add in redundancy such that certain roles or niches are repeated within the ecosystem (for example, there may be many decomposers capable of breaking down lignin). Thus it may be possible for some species to disappear from an ecosystem without the ecosystem collapsing. However, some species are more important than others in an ecosystem. If the keystone species are lost, the entire ecosystem undergoes drastic changes (Paine 1966). Unfortunately we often don't know the identity of these keystone species until they are lost and the ecosystem is drastically changed (Golubiewski 2008). In general it is clear that additional biodiversity is a benefit to ecosystems. Given our new understanding of the importance of biodiversity to ecosystem functioning, the global losses of biodiversity become even more frightening and critical to prevent.

Saying that reductions in biodiversity lead to decreases in ecosystem services is such clinical language that it is easy to underestimate how devastating such decreases are. It is helpful to put the effects of those declines in ecosystem services in human terms. It has been estimated that declines in ecosystem services such as pollination, pest control, water cycles and nutrient cycles may result in a 25 percent reduction in global food production by 2050 (Nellerman and Corcoran 2010). The fact that the projected decline in food production will occur at a time when the human population is projected to increase by at least two billion people, from seven billion today to nine billion in 2050, is sobering to say the least. At this moment in time the loss of ecosystem services such as flood-water retention and buffering coastal areas from damaging storms is estimated to contribute to 270 million people per year being affected by natural disasters, with about 124,000 additional people being killed by those disasters. Eighty-five percent of the people affected by the increase in natural disasters live in Asia (Nellerman and Corcoran 2010). The July and August 2010 monsoon flood events in Pakistan, which at one point covered about one-fifth of Pakistan's land with flood waters, impacted at least 20 million people (about 10 percent of Pakistan's total population) and left an estimated two million people homeless (Goodwin 2010). That huge flood is widely thought to have been exacerbated by deforestation and thus the loss of ecosystem services such as flood prevention. Continued depletion of global ecosystem services will almost certainly lead to additional disasters similar in scale to the tragic flooding in Pakistan.

Despite recent controversies about the reality of global climate change in the popular media, the evidence that global climate change is occurring is incontrovertible and accepted by the vast majority of scientists. As Parmesan (2006, p. 639) succinctly states:

> Direct impacts of anthropogenic climate change have been documented on every continent, in every ocean and in most major taxonomic groups.... Twentieth-Century anthropogenic global warming has already affected the Earth's biota.

In Chapter 4 I will explore the effects of global climate change on ecological restoration in detail, but I will note here that global climate change will greatly complicate both our ability to conduct good ecological restoration due to increasing difficulty in matching historical ecosystem conditions and our understanding of what good ecological restoration even is (Harris *et al.* 2006).

The spread of invasive species is also causing declines in biodiversity and changes to ecosystems. There are positive feedbacks between invasive species, changes in land use and global climate change, which allow invasive species to spread and proliferate even more rapidly. Many invasive species are tolerant of human land-use patterns and disturbance regimes and thus are well adapted to increasingly human-dominated landscapes. Because many native species are already under stress and experiencing population decline due to habitat modification, the added stress of climate change further weakens their populations, allowing invasive species adapted to climates more similar to the newly developing climate regimes to win in competitive interactions with native species.

The scale and success of invasive species appears to be unprecedented at the global scale (Ricciardi 2007). The positive feedbacks that favor rapid spread and increases in invasive species populations lead to a phenomenon that Ricciardi (2007) terms "invasional meltdown," in which ecosystems can be transformed rapidly from native ecosystems to novel ecosystems (Hobbs *et al.* 2006). The arrival of invasive species and development of novel ecosystems is another set of factors that both complicate our ability to do ecological restoration and our understanding of what good ecological restoration should be (Hobbs *et al.* 2009; Allison 2011). Our typical strategy of dealing with invasive species via three different measures (usually applied in sequence) – (1) prevention of establishment, (2) if prevention fails then eradication and, finally, (3) if eradication fails then control – clearly is poorly suited to handling the current spread of invasive species (Hulme 2006). Prevention and eradication are failing at both local and global scales and it is becoming doubtful that we can even control invasive species in many cases (Hobbs *et al.* 2009). I will return to the challenges posed by invasive species and novel ecosystems in more detail in Chapter 4, but at this point it is obvious that ecological restorationists have many challenges due to climate change and invasive species ahead of them in the twenty-first century.

There was a time, not so very long ago, when ecological restoration was thought to be a dubious practice by many conservationists. Sometimes restoration was seen to be completely antithetical to the practice of conservation and the preservation of biodiversity and wild ecosystems (Katz (1992) provides perhaps the purest example of how skeptical many people were about restoration, and Higgs (2003) provides an excellent summary of both sides of the issue).

Skepticism about ecological restoration rested on two major points. First, could restored ecosystems ever truly become wild in the same way that their model ecosystems were wild? Second, success at restoration might well promote hubris that would lead us to develop/destroy ecosystems, thinking we could easily repair them. However, even as Katz was voicing his objections to ecological restoration in the early 1990s, several conservationists recognized that the rate of habitat destruction and loss of biodiversity was so great, and so many ecosystems were decimated if not completely destroyed, that preservation of ecosystems would never be enough to save biodiversity and ecosystem functioning. It was apparent more than 20 years ago that ecological restoration would have to be an essential part of our toolkit for maintaining the earth's wild and natural resources and ecosystems (Leopold 1939; Turner 1987; Janzen 1988; Baldwin *et al.* 1994; Young 2000). Since then it has become even more obvious that ecological restoration will be one of the most important tools, if not *the* most important tool, used to preserve biodiversity and ecosystem functioning in the twenty-first century and beyond (Hobbs and Harris 2001; Harris *et al.* 2006; van Andel and Aronson 2006; Clewell and Aronson 2007; Hobbs *et al.* 2009; Mooney 2010; Nellerman and Corcoran 2010; Rands *et al.* 2010).

The rapid transformation of the landscape, loss of biodiversity, usually negative changes to ecosystem structure and functioning, on-going and future global climate change, spread and success of invasive species, and environmental damage due to many pollutants all provide reasons for us to engage in ecological restoration. But we will not engage in ecological restoration unless we choose to do so. That choice has to be an active choice that matches our desires and needs, just like the choices to engage in agriculture to produce food, to develop the land to build houses and towns, and to mine for coal to supply energy are all active choices that meet human needs and desires. This leads to the question: Why would we choose to conduct ecological restoration and how does the choice to do ecological restoration fit into our usual set of priorities? As we shall see, there are many reasons to engage in ecological restoration and different people place different degrees of emphasis on particular reasons. Similarly, people may make very different assessments of the need for ecological restoration and thus the priority it should receive when determining how it should fit into various land-management plans.

I recently attended the Mid-Shires Sheepdog Championships in Drayton Parslow, Buckinghamshire. While there I chatted with several of the farmers and shepherds running their dogs in the championship trials. One fellow in particular was curious about what brought an American to rural England to live and work. I explained that I was there conducting research into ecological restoration. As soon as he heard the phrase "ecological restoration" he scoffed about ecologists and the idea of restoration. To him ecological restoration meant taking good land out of production for ecological reasons, and that made no sense when "so many people around the world are hungry." I tried to explain that good ecological restoration would be beneficial to agriculture, but he brushed that aside, saying it was crazy the way the government was paying people to take land out of production and have it sit idle, and complained about government restrictions to hedge trimming that made farming harder. I thought about engaging him in a discussion about how we produce more than enough food to feed the world's population, but various government and market policies prevent food from reaching people who need it, but I could tell I wouldn't make much progress with him. So I decided to let it go and just enjoy watching the border collies work instead.

We also know that land management choices may sometimes be made for reasons that make sense in the short term, but which do not fully recognize the long-term consequences of that choice or the long-term needs of people and the environment. Indeed, we often seem to fall into patterns of behavior that in the long run lead to particular negative results, without questioning why we are doing so:

> Every uneventful day that passes ... reinforces a steadily growing false sense of confidence that everything is all right – that I, we, my group must be OK because the way we did things today resulted in no adverse consequence.
>
> (Snook 2002, pp. 199–200)

The choices we make each day to engage in a particular behavior, whether it is to cut down some trees on a hillside, to pull out a hedgerow to allow for the planting of a few extra rows of crops, to surface-mine for coal to burn in an electricity-generating power plant or just to drive our car down to the grocer rather than walking, all seem rational at the time. After all, we have done these things in the past and nothing bad happened. So it seems perfectly rational to continue with those behaviors in the future because we do not see any negative consequences until it is too late, until some kind of disaster happens such as a huge flood because there are no longer any forests to absorb rainwater and slow its passage into the rivers. Garrett Hardin pointed out many years ago that our environmental problems often result from treating the environment as a commons that no one owns but everyone has access to and equal incentive to use for personal benefit (Hardin 1968). Neoclassical economists have long assumed that each individual human, thought of as a consumer and often modeled as *Homo economicus*, makes economic choices rationally. But to make rational choices we essentially need to have perfect knowledge of the costs and benefits, including the long-term outcomes, of our choices (Farber *et al.* 2002). Far too often we do not have perfect knowledge, so we cannot easily envision what the long-term consequences of our actions will be. Thus choices that seem rational at the time, because they have always worked before, are not rational in the long run. We too often make choices that lead to environmental degradation, and that degradation leads to the need for ecological restoration.

How are choices made?

We know people are not always rational actors and sometimes, probably often, make choices that make sense in the short term even though in the long term they might be better off with different choices. How, then, do we make choices about ecological restoration? And how do we know if the choices we make today will really be the best choices in the long run? The decision to introduce autumn olive to North America in the 1950s in order to provide native animals with a plant that produced both abundant fruits and shelter made perfect sense at the time. However, autumn olive has become an invasive pest species, often out-competing native plants and changing ecosystem properties (Allison 2011). I constantly worry that the choices I make today will turn out to be as ill-advised as planting autumn olive, but because ecological restoration is a crisis discipline, rather like conservation biology, we are forced to make choices without perfect knowledge of all the consequences of that choice (Soulé 1986). We have various guidelines that have been developed that should help us to avoid making outrageous blunders, but there is still an element of risk involved with all choices, and thus as Eric Higgs has pointed out, we need to make our choices with humility and a fear of hubris (Higgs 2003).

Given the diverse backgrounds of people all across the globe involved in ecological restoration and the varied traditions of ecological restoration that developed at different times and places, it is not surprising that there are many different reasons given for how choices are made about why, how and when to carry out ecological restoration. Joan Ehrenfeld (2000) claimed that the reasons people give for engaging in ecological restoration can be placed into four rather broad (and to my mind not entirely mutually exclusive) categories.

One set of reasons are largely related to conservation, and focus on restoring particular species and ecological communities to a specific site.

Another set of reasons relate to land and ecosystem management. In this instance, people are most interested in ensuring that the landscape continues to function in particular ways that are desirable to us and that protect resources like forests, grazing land and water resources. Here the focus is often on restoration at the landscape or watershed level. Closely related is the idea that ecological restoration should primarily be done in order to provide particular ecosystem functions such as flood control, nutrient cycling and maintaining the normal hydrologic cycles. While this approach is similar to the goals of landscape management, the subtle difference is that here the desire to preserve ecosystem functions and services is made more explicit and the goal is less to maintain a particular land use than it is to simply ensure that ecosystems continue to function in a way that benefits us and the greater ecosystem. Many people primarily interested in restoring ecosystem function and services originally came from backgrounds in wetland restoration, where the choice to engage in restoration was often driven by practical considerations like flood control and water purification.

Finally, in some cases, where the land and ecosystems have been severely damaged by mining or industrial processes resulting in derelict land or brownfield sites, the goal is simply to restore some kind of ecosystem that provides improved function compared to the derelict land. There may be no attempt at all to restore to the original, historical ecosystem or landscape. Ehrenfeld's framework provides a good starting point for considering the range of reasons why people choose to restore ecosystems, but I would add that sometimes people feel a moral obligation to do ecological restoration. Restorations that produce economic and social values are becoming increasingly important reasons for people as they plan projects, and such considerations will only become more important as the human population grows and the need for ecological restoration increases during the twenty-first century.

Restoration of destroyed and/or lost natural ecosystems

For many practitioners of ecological restoration, especially North American and Australian practitioners (see Chapter 6 for data and discussion), the main reason to engage in ecological restoration is to restore a damaged ecosystem to its previously existing, undegraded, historical condition. As noted in Chapter 2, at the very beginning of the modern development of ecological restoration in North America, such as the early restoration projects at Vassar College and the University of Wisconsin Arboretum, the primary motivation was to return to the plant communities that had been present prior to changes in land use and disturbance of those ecosystems.

In 1997, when Eric Higgs wrote a description of what constituted good ecological restoration, his first criterion was that good restoration has to have a historical component. For Higgs, at that time, good restoration had to be based on what was present on the site prior to disturbance and as much as possible must attempt to match those previously existing conditions. Although Higgs probably would not be so insistent about making historical conditions the most important aspect of good restoration today, he nevertheless continues to stress that good restoration has to take into account the pre-disturbance ecosystem, and he feels that restoration must consider past conditions or in some way it is not a complete restoration (Higgs and Hobbs 2010). The very thing that makes ecological restoration unique among various conservation and land-management practices is the fact that in ecological restoration there is a strong focus on re-establishing the undegraded condition to an ecosystem, sometimes even attempting to achieve the pre-human condition (Munro 2006).

Although I tend to think the insistence that ecological restoration should be based on a return to historical conditions is primarily a feature of restorations in places like North America and Australia, where there is an obvious change in land management associated with the arrival of European peoples, there is strong interest in restorations that reflect historical conditions in Europe as well (Box 1996; van Andel and Aronson 2006). In his seminal book, *The History of the Countryside*, Oliver Rackham (1986, p. xiv) claims:

> It increasingly seems likely that, at least since the Iron Age, every inch of the British Isles has either belonged to somebody or has been expressly set aside for communal use. Not just main roads, but wide areas of fields and lanes are Roman (or earlier) antiquities, and survived the Dark Ages almost intact.

For Rackham, changes to the historical landscape result in four kinds of losses: (1) the loss of beauty; (2) the loss of freedom as commonly held spaces and paths are either destroyed or privatized; (3) the loss of historical vegetation and wildlife; and (4) the loss of meaning, which is his main concern. He sees the landscape as a record of the roots and history of people in the British Isles and claims each piece of the landscape has something important to tell us. Thus, it is important that we maintain and restore long-standing, traditional land-use practices such as grazing, coppicing, pollarding, etc. I should note that in many ways Rackham is rather pessimistic about what restoration can achieve. He claims that ancient woods can never be recreated or restored and that even restoring something simple like ancient grasslands would take at least 200 years. But he clearly sees restoring old land-management techniques as key to maintaining the landscape, and similar restoration of traditional management practices is increasingly of interest across much of Europe, where land uses are at least as ancient as in the British Isles (van Andel and Aronson 2006).

When I was at the 2010 SER European Conference in Avignon, I had a good chat one evening with the eminent Dutch restoration ecologist Ab Grootjans. We talked at some length about the reasons why people want to restore landscapes, and Ab said that he felt that for many restorationists, especially those over the age of 50, restoration is about recreating the landscapes of their childhood. For them, restoration is in many ways an attempt to recreate an idealized, perhaps simpler and cleaner past. I have noticed this desire to return to past conditions that continue to exist within living memory when talking with several lay restorationists.

The desire to restore to previously existing, historical conditions has, almost from the beginning of modern ecological restoration, led to many questions about the feasibility of actually attaining historical conditions. One of the major questions is what historical or pre-historic period we should focus on. In some places like the Americas and Australia, the arrival of Europeans led to obvious, large-scale changes in land management that make the time just before European arrival seem to be a definite reference model. At other locations, especially common on the Pacific Ocean islands like the Hawaiian Islands or New Zealand, the arrival of Europeans was preceded by only a few hundred years by the arrival of Polynesian peoples (a very brief period of time in evolutionary terms), so it might be desirable to restore to pre-human arrival conditions if possible on those islands.

Case study: River Shep restoration at Fowlmere Nature Reserve, Cambridgeshire, UK

In February 2011 I visited the River Shep restoration project at Fowlmere Nature Reserve with an ecological restoration class from Cranfield University. Fowlmere is a small nature reserve owned by the Royal Society for the Protection of Birds (Figure 3.1). The River Shep is a small, low-gradient chalk stream with its headwaters at springs located within Fowlmere. The River Shep suffered significant degradation over the past 30 years due to excavation of gravel from the stream bed. As a result the river channel was deeper than in the historical past, the stream banks were wider, the channel was full of silt and there had been a large decline in the native biodiversity of the stream. In particular, brown trout (*Salmo trutta*), water crowfoot (*Ranunculus penicillatus*) and water starwort (*Callitriche* sp.) had decreased in abundance.

Our host and guide, Rob Mungovan of the South Cambridge District Council, had lived in the area and known the River Shep all of his life. He had seen the stream change from a clear, shallow chalk stream that supported trout to a deeper, silty, muddy stream with few fish. He had also heard many stories about what the stream was like in the past and had seen other chalk streams that were in better shape, so he was determined to restore the River Shep to its former status – at least the section that lay within Fowlmere's boundaries. Rob began working on the River Shep in 1997, sometimes working alone and sometimes with volunteers, trying to stabilize the stream banks and remove excess growth of water parsnip (*Berula*

Figure 3.1 The River Shep restoration project is located within the Fowlmere Nature Reserve, a Royal Society for the Protection of Birds site in Cambridgeshire, UK.

Figure 3.2 The restored section of the River Shep looks like a classic chalk stream – fairly shallow (usually 40 cm or less in depth), with clear water, a bed of chalk gravel and patches of water crowfoot and water starwort growing on the bed.

erecta) that was collecting silt and slowing stream flow. By 2005 it was obvious to Rob that he needed additional help both in terms of expertise and funding in order to complete the restoration. He developed a restoration plan in consultation with the River Research Centre at Cranfield University and received a £14,000 grant to enable him to hire equipment and purchase supplies.

The level of the stream channel was raised by adding in chalk gravel. The stream was narrowed by placing faggots of ash and hazel along the stream margins. Riffles were constructed in a few sections. Rob was extremely fastidious in planning and carrying out the work. He carefully selected gravel of the same size and color as was previously found in the stream bed so that it matched the historical condition. He took many measurements to ensure the stream was as shallow and narrow as before (he thinks it isn't quite shallow enough yet, but it is functioning correctly so he is happy about that).

By 2009 the restoration work was completed and the stream was transformed. Today it supports increasing populations of brown trout, water crowfoot, water starwort and also bullheads (*Cottus gobbius*). It is also starting to support more diversity and abundance of the stream invertebrates that are prey for the trout.

As we walked along the stream I was impressed by the crystal clarity of the water, the trout we startled in riffles and the twinkle in Rob's eye as he told us about the restoration. He knew every centimeter of the restoration and was able to show us where trout had been shifting gravel in the riffles to form spawning redds. He pointed out places where he left fallen logs lying in the stream and where he removed excess water parsnip. He wanted the stream to have considerable physical diversity to help support biological diversity. But mostly it was obvious that he just loved the River Shep – both the stream he remembered from his boyhood and the stream reborn from his and many volunteers' and assistants' efforts. The restored section flowing through Fowlmere is only about 600 meters long and ends abruptly where the stream flows out of Fowlmere into an agricultural field. In the field the stream is shallow and silty due to erosion from the surrounding banks. But within the 600 meters of the headwaters, a bright, clean, diverse and healthy stream has been reborn due to a remembrance of its past condition, the use of modern restoration techniques and the vision and dedicated work of one person.

The logical extension of Katz's (1992) argument that human activity renders landscapes unnatural would be to try to restore to conditions that existed prior to any human activity at all in all ecosystems. However, in many cases that would mean attempting to restore to conditions that existed so far in the past that we have only sketchy and slim data about what the ecosystems existing at that time were like. Or it might cause us to attempt restoration to conditions that are physically impossible to maintain today (would restoring pre-human northern Eurasia and northern America mean restoring glacial ecosystems?).

Even so, some people have claimed that pre-human conditions should be our goal. Tim Flannery has suggested that in order to restore some aspects of Australian ecology, we should aim for conditions that existed prior to the arrival of Aboriginal peoples about 60,000 years ago (Flannery 1994, cited in Ridder 2007). Brian Czech (2004) has claimed that an objective time period to select for pre-restoration conditions would be between 800 and 1800 CE. In his view, that would incorporate a 1,000-year window that includes both warm and cool periods and also predates the rise of industry. For Czech the truly damaging human activity is human economic activity, and economic activity reached its on-going zenith with the development of the Industrial Age from 1800 onwards. However, Czech's time frame seems somewhat arbitrary because we know that many landscapes (much of Europe (especially the Mediterranean ecosystems), China and India) had been highly modified by human agricultural activity long before industrialization. So is industrialization really the point at which ecosystems underwent drastic human-driven change? In either case, it is clearly not a simple matter to identify an objective time period that represents the pre-disturbance, restoration target.

In some ecosystems, like the Midwestern American tallgrass prairie, where almost all tallgrass prairie was destroyed by conversion to other land uses (primarily agriculture and urbanization) before anyone had a chance to carefully study the ecosystem, the first question is to ask how well we even know what the pre-disturbance conditions were like. In such cases our best and only window into the past is often to examine the few remaining remnants of a once much more extensive ecosystem. Those remnants serve as a model reference against which we can measure our restorations. But it is obvious that remnants do not provide a perfectly clear window. We must ask: Why did these remnants survive? In the case of tallgrass prairie, where so much prairie was plowed under in order to plant crops, is it possible that the remnants were somehow less desirable for agriculture and thus not developed because they were different from more typical prairie? Even if they were similar to the original tallgrass prairie, we know from studies of island biogeography and the creation of nature reserves that once an ecosystem is cut off from an originally continuous ecosystem and surrounded by dissimilar ecosystems, the now remnant ecosystem almost always undergoes changes due to loss of species and physical changes to the ecosystem due to edge effects and the loss of continuity with a larger whole (Lovejoy 1985; Newmark 1986). It is almost certain that remnants of once-widespread ecosystems have suffered similar changes after they became isolated from their original surroundings.

We also know that even without human-caused disturbances, ecosystems are dynamic, constantly changing; they are not static, fixed entities (Botkin 1990). Thus, even if no human disturbance had ever occurred, the original ecosystem would have undergone changes from the historical conditions we are trying to match, although it is impossible for us to know what those changes would have been. Such questions have led to the criticism that an overly slavish devotion to a certain set of historical conditions will lead to restorations being little more than snapshots from the past or dry, dead museum pieces that do not contain the vital spark of change and growth typical of living systems (Dunwiddie 1992). Thus, even if historical conditions and the pre-disturbance ecosystem are the models when we begin planning ecological restoration, the planning must call for restorations that are dynamic and have the possibility of continual change. Restored ecosystems must be free (and hopefully large enough) to allow for the full range of evolutionary processes and ecological change typical of undisturbed ecosystems – in other words, they must be returned to their "historical trajectory" even if not to their exact historical past (Society for Ecological Restoration Science and Policy Working Group 2004; Higgs 2003). Another way of framing this issue is to manage restorations to encompass the historical range of variation (HRV), because we know that in the past ecosystems were variable. If we can somehow determine how variable they were in the past, we can use that degree of variability to set limits to the amount of variation we think is acceptable in the restoration today (Landres *et al.* 1999).

One aspect of choosing to restore ecosystems to a model based on the undisturbed, historical past that is usually not directly stated, is that for many people choosing a return to past conditions allows us not to make a choice at all, or to pretend we are not making a choice when in fact we are choosing. If we return to the past, we are merely reading the book nature wrote and returning to the point where we lost the plot. In this model we are simply removing our presence, but the identity, structure and function of the ecosystem is nature's choice, not ours. Thus we avoid the problems that arise when we attempt to impose our will on the environment during the process of restoration (i.e., the problems of authenticity that bother Katz (1992)). There are a few authors who explicitly make the claim that nature knows best and will make the correct decisions for us. Steve Packard has said that "Nature will tell us if we've got it right" (Helford 1999). In fact, he has written that his early approach when restoring Midwestern oak savannas was to unleash "natural forces" like fire and the fire would do the work of "deciding" where prairie ended and savanna began:

> "Let the fire decide," became our motto. That was the natural scheme, that's what we wanted.
>
> (Packard 1988, p. 14)

Later Packard learned it was not so simple and that considerable human intervention was necessary to produce the ecosystem he desired to recreate on former savanna sites. Dan Janzen, who is usually very pragmatic about what ecological restoration can accomplish and why it is necessary, wrote:

> How does one know what habitat to restore? The simplest thing is to let the organisms decide...
>
> (Janzen 1988, p. 243)

Many restorationists claim that by restoring to undisturbed, historical conditions they are not really putting a new ecosystem into place; instead their work allows them to discover and thus show to all people the authentic natural ecosystem that should exist there. At times the claims of returning to the historical and, in their view, true, correct ecosystem for a particular site causes restorationists to adopt a tone of moral superiority that does not sit well with other people, whether they are fellow conservationists or the general public (Helford 1999).

Ecological restoration must have some basis in the undisturbed, previous ecosystem, but we must recognize the limitations of that model for restoration and be sure that what we are attempting to do is attainable. Ultimately, even the most carefully planned restoration will never be able to exactly recreate what was lost due to disturbance, but that should not dissuade us from using the past or reference ecosystems as a model for our work. Restoration that honors the past and allows for dynamic ecologies and evolutionary change will have the spark that defines living systems and that is the essence of "wildness" – those aspects of nature that are beyond human control (Ridder 2007). It is essential that we recognize that there are many things in any restoration that we can't control, but we can set the stage in a way that returns a site to a condition more desirable than its degraded condition, and that is a valid choice for us to make.

Restoration and maintenance of biodiversity

For many restorationists and restoration ecologists, the main goal of ecological restoration is to restore and maintain both local and global biodiversity (Janzen 1988; Dobson *et al.* 1997; Clewell and Aronson 2006; Aronson *et al.* 2007; Mooney 2010). Biodiversity can only exist in an ecosystem and is a fundamental component of ecosystem functioning and services, but it is important to note that for many people the goal of restoring biodiversity is a good enough reason in itself, regardless of whether the added biodiversity results in the restoration of historical ecosystems or contributes to ecosystem services. This may seem a subtle point, but as we try to understand different people's motivations for being involved with ecological restoration it is important to acknowledge that for some the restoration of biodiversity for its own sake is sufficient motivation to engage in ecological restoration (Aronson *et al.* 2007).

We know that biodiversity is under siege and rapidly decreasing due to human population growth and increasing domination of the earth's resources (Vitousek *et al.* 1997; Sanderson *et al.* 2002; Millennium Ecosystem Assessment 2005; Mooney 2010; Nellerman and Corcoran 2010). Therefore, using restoration primarily to restore and promote biodiversity at all of its levels (most commonly understood as a combination of genetic diversity, taxonomic groups of species, functional groups of species and ecosystem diversity) and to the greatest extent possible is more than enough justification for many ecological restoration projects. A focus on restoring biodiversity is not restricted to scientists, governmental and non-governmental organizations. More and more individual restoration enthusiasts, such as prairie plant aficionado and master seed collector Jay Stacy at Nachusa Grassland, are also highly motivated by the idea of promoting biodiversity as the primary benefit of ecological restoration.

What is curious is that an initial focus on biodiversity can lead restorationists down a few different paths and to different conclusions about the best way to achieve their goals. One path is that followed by Jay Stacy, which leads to an almost exclusive and somewhat obsessive pursuit of producing restorations that contain every species found within a fairly well-defined local area. Restorationists following this path are especially interested in small, local populations and communities and preserving the adaptations that have evolved at or near a particular site. Although this path may develop among restorationists working in relatively species-poor ecosystems like coastal salt marshes, where there may be fewer than 15 native plant species, it is more common to see restorationists fixated on species diversity when they are working in species-rich ecosystems that frequently have large numbers (often hundreds) of rare or uncommon species – such as prairies, savannas and chalk grasslands.

For some restorationists the key component of biodiversity is genetic diversity. This is partly because maintaining all levels of genetic diversity in a species may well capture a broader range of diversity than simply maintaining biodiversity at the level of species. Many species, especially those that live in somewhat isolated populations with poor dispersal between populations, will almost certainly have a lot of genetic variation that has arisen due to adaptation to local conditions among those populations. Just preserving the species (where saving one population would be enough to maintain the species) would then lose the true range of genetic diversity that exists in the species. Another concern that arises with losses of genetic diversity is that as global and local climates change, local conditions will change as well. If a population is well adapted to the set of conditions that exist today, it may be poorly adapted to future local environments that may arise in the next 50–100 years. Therefore, it is vital to preserve as wide a range of genetic variation as possible within the species so that there is a greater chance the species will have the genetic variation needed to adapt to climate change (McKay *et al.* 2005). In this instance, preserving genetic diversity for the sake of increased genetic diversity (and therefore biodiversity) is an insurance policy for the future and will be vital for maintaining species in the future.

At the ecosystem level we are usually interested in both the variety of ecosystems that exist within a landscape and the variety of functional groups of species that exist within each ecosystem. Functional groups are usually first divided into trophic groups – primary producer, primary consumer, decomposer, etc. They are then further subdivided into types of each trophic group – so for primary producers we may divide them into woody plants (perhaps broken down into trees and shrubs), C3 grasses, C4 grasses, leguminous herbs, other herbs and algae, or whatever system makes the most sense for the ecosystem under consideration. During restoration we would want to ensure that all functional groups originally found in the undisturbed ecosystem are present post-restoration and that they are carrying out the same functions as before (Naeem 2006).

Finally, restorationists mainly interested in biodiversity may well ask, isn't biodiversity simply biodiversity? If we are interested in preserving biodiversity, then shouldn't all species regardless of place of origin count? Most restorationists are mainly concerned with restoring species native to the site at which they are working. Non-native species are usually viewed skeptically if not with outright disdain and subjected to eradication policies, because non-native species are perceived as not belonging and because they have the potential to become dangerous invasive species that out-compete native species and greatly alter ecosystem properties (Allison 2011). As anyone who has worked in ecological restoration for very long realizes, complete eradication of non-native species is virtually impossible (Ewel and Putz 2004; Hobbs *et al.* 2006, 2009; Ricciardi 2007; Allison 2011). In fact, I will go out on a limb and claim that completely eradicating all invasive species from ecological restorations is impossible. My students and I spend hundreds of person-hours every year trying to remove invasive species from Green Oaks. I often find myself developing a dark and deep antipathy towards the invasive species, especially the woody and thorny species like black locust and autumn olive that are so adept at resprouting from their roots. To my chagrin, I have noticed my own students also developing a strong dislike, even hatred, of certain invasive species (an attitude I am afraid they have learned at least in part from me). Such responses to invasive species are not unusual among restorationists, many of whom use the language of warfare to describe their attempts to "do battle" with non-native species in order to "vanquish" the "invasive menace" (Gobster and Hull 2000).

It is almost certainly better to adopt a more balanced attitude towards non-native species (Ewel and Putz 2004). If the non-native species really are here to stay in our ecosystems, we need to recognize two things: (1) even if it were possible to remove all of them from restorations the costs to do so would be so high that only a few projects could afford to achieve complete removal; and (2) we may be missing the fact that non-native species are providing benefits to us in the restoration process. Non-native species may serve a vital function by making positive modifications to the environment during restoration and acting as nurse species helping native species become established (Ewel and Putz 2004). Certainly, as the global climate continues to change and as human land use continues to expand, the preservation and restoration of any amount of wild, non-domesticated biodiversity will be valuable regardless of its place of origin (Hobbs *et al.* 2006, 2009).

I will add the caveat that welcoming or at least accepting non-native species in our restorations is problematic because often we don't know the exact effects that non-native species will have on any given ecosystem and because some non-native species appear to be benign members of an ecosystem for decades before becoming a problem due to some additional environmental change (Elton 1958). As with all things in ecological restoration, we must guard against hubris and act with humility regarding our understanding of the impacts of non-native species (Higgs and Hobbs 2010).

I don't want to end on a negative note with respect to practicing ecological restoration with a primary goal of restoring biodiversity. The loss of biodiversity is more than just troubling; it is catastrophic (Mooney 2010). Biodiversity must be restored because it is vital for maintaining ecosystem structure and function. But, there is much more to restoring biodiversity than protecting ecosystem services. The simple joy of seeing the earth's multitudes of species and the marvelous adaptations of those species is the reason so many restorationists focus on restoring biodiversity. There is something delightful about bending down and being surprised by a rare species you haven't seen before or hadn't seen in many years. I had heard there was a relict population of an Illinois state endangered species, the four-toed salamander *Hemidactylium scutatum* in the woods at Green Oaks. So for several years I dutifully went out looking for it, turning over logs and peering underneath them, getting down on my hands and knees and gently probing the leaf litter. And then one autumn day I found one, just one, a very small, somewhat drab reddish-brown salamander, perhaps 8 cm long, crouched in the moist soil. On close inspection it had just four toes on its hind legs. "Wow!" I said to myself, "I finally found you." It was like reaching the prize at the end of an arduous scavenger hunt. I hurried off to e-mail my former student Mathys Meyer the good news because he was the one that had originally set me off on this quest (Meyer *et al.* 2002).

Another year, late spring, found me in Lost Meadow doing my annual sampling program – laying out transects, counting plants (Allison 2002). Out of the corner of my eye I caught sight of a beacon of reddish-orange. Could it be my old friend Indian paintbrush (*Castilleja coccinea*), someone I hadn't seen for about ten years in this meadow? I walked over and bent down, and indeed it was. I couldn't help but smile and say, "Welcome back, I've missed you." Yes, biodiversity is well worth preserving just because of the joys and surprises it brings to us unbidden, but always possible.

Restoration of severely degraded sites

From the beginning of the modern practice of ecological restoration to the present, by far the most challenging sites for restoration are those where the site to be restored is so completely degraded as to be effectively destroyed as a functioning ecosystem (Bradshaw and Chadwick 1980; Cairns 1980). Severely degraded sites typically lack any topsoil or vegetation, or if soil is present the soil has been so badly contaminated by various pollutants that it can no longer support the growth of plants. Many of these sites were created by surface mining for various minerals or products like clay soils or gravel. The pits left behind by mining have severely altered topography and hydrology compared to what was there before mining occurred. Surface and shaft mining can both create large piles of mine spoils that are frequently toxic because they contain high concentrations of various metals or have a pH that is too acidic or basic to support plant growth. Because degraded mine sites are so obvious and because a considerable amount of effort has been dedicated to restoring them over the past 60–70 years, we often think of mine sites as being the most prevalent type of severely degraded ecosystem. However, there are many other types of severely degraded ecosystems present today. Direct input of pollutants such as heavy metals by aerial deposition (Winterhalder 1984) and run-off (Wu 2004) has resulted in contamination and destruction of ecosystems. Soils with extremely high levels of salinity have been created by agricultural practices that cause changes in the water table and normal hydrological function (Hobbs *et al.* 2003). The abandonment of urban industrial sites has left behind brownfield sites, which usually have derelict buildings and soils contaminated by various hazardous substances such as heavy metals and organic chemicals (Dorsey 2003). Perhaps because brownfields are usually located in urban environments (although there are rural brownfields too) they have only recently started to attract the attention of many restorationists, but brownfields are unfortunately a very common part of our landscape today. It was estimated that there were 450,000–600,000 brownfield sites in the United States in 1999 (Dorsey 2003). These sites ranged in size from small (such as abandoned neighborhood gasoline stations) to quite large former industrial complexes. Other industrialized nations are also likely to have large numbers of brownfield sites in need of restoration.

The main goal of restorationists working with such damaged ecosystems has been primarily to take an ecosystem that is no longer functioning (many have bare soils and thus lack primary production and biological nutrient cycling, and suffer from high rates of erosion) and restore it to an ecosystem that has some kind of natural ecosystem functioning and that will be able to maintain itself over time (Bradshaw and Chadwick 1980; Cairns 1980). These sites require restoration because when left alone they are often hazardous due to the pollutants present, the potential for pollutants to flow out of the damaged ecosystem and thus poison other ecosystems, the high rates of erosion occurring at the site and the extremely slow rates of natural recovery if they are simply abandoned and left to recover by ecological succession (Bradshaw 1984). If left alone, these sites also remain an eyesore, a literal blight on the landscape, and in their unattractive state they are thought to be useless to humans.

Because restorationists usually have a strong desire, and often a legal mandate, to turn these severely degraded ecosystems into some kind of ecosystem that ameliorates the negative conditions that exist at the site and produce a viable ecosystem with positive benefits to us and the greater global ecosystem, they often have not been as interested in recreating past historical conditions on these sites as they have been in simply establishing a successful ecosystem (Bradshaw and Chadwick 1980; Cairns 1980). For instance, an area that was forested prior to surface mining may be restored to grassland or redeveloped as agricultural land or turned into a park and golf course if those options are easier, less expensive than a historical restoration and acceptable to the stakeholders involved.

Restorations that produce an ecosystem that is different than the predisturbance ecosystem are sometimes highly regarded by people living locally and even much farther afield. Butchart Gardens was created on a former limestone quarry located on the outskirts of Victoria, British

Columbia. It opened in 1921 as a pleasure and botanical garden and has been a popular destination ever since. The Eden Project in Cornwall is located on a former china clay pit mine (Figure 3.3). It was designed as a garden that would display and help preserve examples of both local and global diversity, with specially designed domes sheltering rainforest and Mediterranean ecosystems. It also includes areas where ecological restoration has returned the site to near pre-disturbance conditions and that help educate visitors about the benefits of ecological restoration. The Eden Project has also been popular with visitors ever since it opened in 2001. Neither Butchart Garden nor the Eden Project fit the usual definition of an ecological restoration project, but both have resulted in ecosystems (admittedly highly managed ecosystems) that provide a great amount of ecological, social, cultural and educational value, especially when compared to what would be present if they had remained as abandoned mining sites.

Because restorationists working with severely degraded ecosystems have usually been more interested in restoring some kind of positive ecological benefit to the disturbed site than in returning that site to the historical,

Figure 3.3 The Eden Project is located in an old china clay pit mine in Bodelva, Cornwall, UK.

Note
The Eden Project is perhaps best known for its large, geodesic-dome-enclosed rainforest and Mediterranean biome gardens. This photograph shows the many demonstration gardens that line the old walls of the pit mine. These gardens show plants from around the world and also display many different plants that provide food, fiber and medicine to humans.

pre-disturbance condition, they have focused on restoring ecosystem functions and services to the site even before those terms were used to describe their work. At this point it is worth stopping to consider the difference between ecosystem functions and ecosystem services, even though the terms are frequently (and unfortunately) used as synonyms. Ecosystems are most commonly conceived as being the living organisms and physical environment that occur at a particular site – so I might talk about a pond ecosystem or meadow ecosystem and I would be referring to all of the organisms and physical factors that make up that pond or meadow. Ecosystem functions result from the combined activities of all components of the ecosystem. There are many ecosystem functions, but among the most frequently discussed are primary production (usually by photosynthesis), energy flow, nutrient cycles and hydrological cycles. Ecosystem services result from the human values that we place on particular ecosystem functions. Thus ecosystem functions are simply the things an ecosystem does, but the services derive from human needs and a desire for what the ecosystem does (DeGroot *et al.* 2002; Farber *et al.* 2002; Naeem 2006; Nellerman and Corcoran 2010).

Restoration of highly degraded ecosystems is a complicated task. It is complicated partly because of the variety of ecosystems that result from severe damage to natural ecosystems. The remains of a copper mine are different from the remains of a coal mine, which are different again from the remains of a china clay mine, which are different again from the remains of selenium run-off, which are different yet again from a brownfield site left behind when an oil refinery shuts down. Of course, on top of that variation in damaged ecosystems needing repair, there is also variation in the original site conditions – a bauxite mine in Western Australia has different original conditions than does a bauxite mine in Guinea. Thus the methods used to repair the damaged ecosystems will vary with type of damage, original ecosystem, local climate and local geology. Many different restoration methods have been attempted around the world. We have learned that even when repairing one ecosystem type in one region of the world – such as restoring Jarrah forest following bauxite mining in Western Australia (Grant *et al.* 2007; Majer *et al.* 2007) or mixed conifer–deciduous forest following aerial deposition of heavy metals near Sudbury, Ontario (Rayfield *et al.* 2005) – that the different methods employed result in different developmental pathways for the restored ecosystem and different outcomes for the restoration – at least in the time span observed so far.

We have also learned that for some ecosystems – such as former agricultural land contaminated and rendered useless by salinization in Western Australia (Hobbs *et al.* 2003) – it may be impossible to restore some highly degraded sites. In those situations we may simply lack the ability to restore the site, or the restoration process may be so expensive and complicated that even though we know how to do the restoration, it is not practical for us to do so. In such cases, we may have to perform triage and restore the sites with the best chance for a successful restoration and abandon the others to recover on their own (Hobbs *et al.* 2003).

Ecosystems that have been damaged and which we do not attempt to restore will undergo some kind of successional process. The main problem from our perspective is that natural ecological succession tends to be slow by the standards of human lifetimes and may not result in an ecosystem that provides services we value. However, there is some evidence that even for severely degraded ecosystems there are times when little or no restoration produces beneficial results (Prach 2001a, 2001b). Tischew and Kirmer (2007) examined the restoration of surface-mined land in the Saxony-Anhalt region of Germany. They found that typical restoration projects there produced flat, level ground that supported a few species of plants (often non-native plants) that tend to do well after human planting. In contrast, they found that for surface-mined sites that had minimal restoration where "natural" measures such as spreading cuttings of vegetation on the ground to slow erosion were used, the resulting restoration was more similar to unmined areas than the actively restored sites. The most important factor in the minimal restorations was the presence of a nearby source of plant propagules or restorationists providing a large number and variety of plant propagules.

The minimal restoration regime resulted in much variation in soil, and topographic and micro-climatic conditions that favored establishment by a highly diverse mixture of native plant species. Colonization by native plants typically took 14–55 years on the minimally restored sites. Even tropical rainforests, once thought to be so delicate that after they were clear-cut they could never recover, are able to recover from disturbances such as clearing and agriculture by natural succession within about 100 years (Chazdon 2008). In fact, across most of Meso-America the forests that exist today are thought to be secondary forests that arose in the past 350 years following the collapse of the Mayan civilization and cessation of large-scale Mayan farming and land clearing (DeClerck *et al.* 2010). Ecological restoration is important in assisting natural succession in tropical rainforests when the soil has been degraded or there is not a nearby source of seeds for establishing a new forest (Chazdon 2008).

These examples of natural succession/restoration in such divergent ecosystems as German forests and tropical rainforests provide us with some hope that over time there will be natural ecological succession and recovery of at least some sites that we do not have the ability to restore ourselves.

The fact that restoration sometimes proceeds more quickly when we do less work should give us pause and force us to ask ourselves how well we really understand the systems we are attempting to restore. Such news definitely should lead us to be cautious and help guard against restorationists developing hubris. But there is another aspect to this dilemma of determining how well we actually understand the systems we are restoring. And that is that, as noted before, ecological restoration is a crisis discipline and as such we often have to make decisions before we know enough to be able to predict outcomes of our efforts with a high degree of precision. Scientists usually want highly significant results (at least at the 95 percent confidence level) that our treatments work the way we think they will before recommending that anyone follow a particular course of action. Certainly that is the way I was trained as a scientist. So we like to test our hypotheses with highly replicated experiments and proper controls that are amenable to statistical analysis. Because the world is a complex place, we often wind up doing multi-factorial experiments that result in a need to do multi-factor analyses, which means we need large numbers of replicates to be confident in our results. Conducting experiments with many factors and replicates requires a lot of time and effort.

Steve Packard (1988) and Robert Cabin (2007, 2011) have both asked whether all the time and effort dedicated to carefully replicated experiments is in fact time that might have been better spent just getting on with the task of restoration. Both think that they have learned as much, if not more, about restoring their systems (Midwestern savannas and Hawaiian tropical dry forests, respectively) simply by doing the work as they have by carefully constructed experiments. Packard, in particular, thinks that if he and his colleagues had waited for the results of experiments to become clear before doing restoration work, he would have accomplished little actual restoration. In fact, he thinks that waiting for the outcome of experiments would have been counter-productive because the savannas he was restoring would have continued to decline and suffer worse degradation during the time it would have taken to do highly controlled, replicated experiments, thus ultimately making restoration more difficult.

This is an attitude I have encountered many times. When I first started working in restoration of tallgrass prairies and I introduced myself as a scientist studying restoration ecology to lay or amateur restorationists, they would frequently roll their eyes or react with a fair amount of impatience as I asked them questions. From their viewpoint my fellow scientists and I were asking questions that they already knew the answers to. They knew how to restore a prairie, so for them conducting the typical field experiment – which is usually fairly small-scale treatment plots (maybe just tens of square meters) and the usual sample size is often just one or at most a few square meters – was a waste of time. For them it made more sense to expend the effort spent on experiments on just doing the work of restoration. My experience has also been that many restoration scientists tend to converse in highly technical jargon at all times and unintentionally either overwhelm their amateur colleagues or, even worse, accidently talk down to them. Scientists are often our own worst enemies when it comes to communicating with the public (Olson 2009). When I first talked to Ted Gilles about the 500 acres of prairie restoration on his Illinois farm, he became visibly nervous and hesitant when talking to me, despite the fact that he has spent 17 years on his restoration and is clearly an expert. I made a note that in the future I should just introduce myself as someone interested in restoration, not as a professor studying restoration. The interaction between scientists and lay practitioners is something that should be explored in more depth and with sensitivity. My experience has been that lay restorationists almost always know what they are doing and produce excellent restorations, even if they can't assign a *p*-value to their results. In some ways ecological restoration remains as much an art as it is a science, and scientists should be careful not to dismiss the excellent work done by the artists in our midst.

Restoration explicitly conducted to improve ecosystem services

Although some restorationists working with severely degraded ecosystems have attempted to restore the sites to the pre-disturbance condition, the difficulties of achieving success by doing so caused many other restorationists to be more concerned with a restoration that produced a healthy ecosystem with some kind of positive value to humans and the rest of the biosphere. This desire to produce something of value led quite naturally to a focus on restoring ecosystem services. Today, ecological restoration with a primary goal of producing valuable ecosystem services is becoming increasingly common (DeGroot *et al.* 2002; Farber *et al.* 2002; Aronson *et al.* 2007; Rey Benayas *et al.* 2009; Nellerman and Corcoran 2010). My analysis (Chapter 6) shows that it is not the most frequently mentioned reason for engaging in ecological restoration, but a discussion of ecosystem services is becoming increasingly important when addressing national and international policy-making bodies (Nellerman and Corcoran 2010).

When I began working in ecological restoration it seemed like almost all restoration projects began with an attempt to restore the ecosystem to its pre-disturbance, historical state (admittedly what I was hearing was a reflection that I was mostly talking to prairie and savanna restorationists and not people doing mine reclamation). But over the past ten years there seems to have been a sea change in our discussions of the reasons for and values of ecological restoration. At first I was puzzled by the increasing number of references to ecosystem services in the context of ecological restoration and the increasing emphasis on spelling out the value of ecosystem services when talking to policy-making and lay audiences. But the more I think about this new emphasis, the less surprising it becomes, because it reflects two new realities facing restorationists. The first is the need to "sell" the necessity for and values of ecological restoration to an increasingly large audience of stakeholders that includes the general public and policy-makers, who may not know much about ecological restoration and/or may not be initially interested in ecological restoration. The second is the mounting evidence that in the twenty-first century, the coming climate change will make our traditional focus on historical restoration increasingly difficult to achieve and perhaps an anachronism. And even though at times I can be a bit romantic in how I think about restoration, I don't fancy becoming a Don Quixote tilting at windmills of climate change in an increasingly futile attempt to turn a small patch of Illinois back to the world of 1820.

The twenty-first century began (at least in terms of discussions of ecological restoration) with Richard Hobbs and Jim Harris claiming that in the new century ecological restoration would become our major tool for addressing environmental problems (Hobbs and Harris 2001). Given that their paper was published in the journal *Restoration Ecology*, they were mostly preaching to the choir. But Wes Jackson, the visionary pioneer of sustainable agriculture, once told me that we need to spend more time preaching to the choir to make sure that we all remain inspired and positive. If ecological restoration is to become our major tool for addressing environmental problems, then it has to provide tangible benefits to people across the world. Ecological restoration attempting to restore historical trajectories and model reference systems will always appeal to some people (like me), but there are also always going to be some people who ask: What is the value of those restorations? And for many of them, a discussion of lost landscapes or preservation of biodiversity for its own sake won't provide much value. Instead they need to see that the restored ecosystem is doing something for them. More than anything, the focus on restoring ecosystem services is a pragmatic approach based on what is possible given changing climates (Hobbs *et al.* 2009) and given the current policy landscape (Nellerman and Corcoran 2010).

There are many ecosystem services and almost as many ways to categorize them. Several authors have divided ecosystem services into four basic groups, although they give the groups slightly different names and definitions (DeGroot *et al.* 2002; Millennium Ecosystem Assessment 2005). The most widely used description of ecosystem services was developed by the Millennium Ecosystem Assessment team (2005). They separated ecosystem services into four broad categories: (1) supporting services (things like primary production, nutrient cycles); (2) provisioning services (production of timber, fish, wild food and fiber, medicines, crops); (3) regulating services (climate regulation, hydrological cycles and water retention, soil characteristics); and (4) cultural services (esthetics, recreation, social value). Ecosystems provide services that are estimated to be worth as much as $72 trillion per year, but two-thirds of global ecosystems are thought to be degraded and thus reduced in their ability to provide ecosystem services (Nellerman and Corcoran 2010). It is vital for us to recognize that for many ecosystems and their services there are critical thresholds, and if those thresholds are crossed, the ecosystem may switch to an alternative state that no longer provides the desired service (Farber *et al.* 2002; Hobbs *et al.* 2009).

Thresholds may depend on the size of the ecosystem, the species composition of the ecosystem, the physical characteristics of the ecosystem or a combination of all of the above (Hobbs *et al.* 2009). Unfortunately, we don't usually know what the threshold is until we have crossed it, and suddenly lose the ecosystem and/or its services. It is also important to realize that the threshold for ecological value of ecosystem services may be different to the threshold for economic value of ecosystem services (Farber *et al.* 2002). Thus, open ocean food webs may continue to function ecologically even when they no longer provide a valued economic product like swordfish or blue-fin tuna. Although determining the economic value of living resources and ecosystem services is controversial (perhaps even highly controversial) (Sagoff 1983, 2009), economic values are increasingly important when presenting the benefits of ecological restoration to the general public and regulatory agencies.

Promoting the economic value and social benefits of ecological restoration is especially important when restorationists work in urban settings and when they are attempting to acquire funding for restoration projects, given the current global economic recession. A meta-analysis of 89 restoration projects located in a wide variety of ecosystems on every continent except Antarctica found that ecological restoration restored 25–44 percent of the original biodiversity and ecosystem services when compared to intact reference ecosystems. There was a strong positive correlation between restored biodiversity and increases in ecosystem services, and that positive correlation was especially strong in tropical ecosystems (Rey Benayas *et al.* 2009). While a return of 25–44 percent of original biodiversity and services due to restoration may sound like a rather small amount, it is still much better than no return at all, and we hope that as our ability to restore ecosystems improves, our return of biodiversity and ecosystem services will also improve. The fact that we are not achieving a 100 percent return of biodiversity and services during restoration is another argument for preserving as many intact, undisturbed ecosystems as possible, but unfortunately the reality is that human disturbance of global ecosystems is already huge and likely to increase in the near future. Ecological restoration will have to be combined with improved land-use practices and land management in order to achieve maximum benefits for humans and the rest of the earth's biota. Organic farming provides 25 percent more ecosystem services than does conventional, modern industrial farming (Nellerman and Corcoran 2010), so ecological restoration combined with a change in agricultural practices has the potential to produce a large benefit.

In urban situations, ecological restoration is not likely to result in returning large areas to natural functioning, but instead will be a vital component of a revitalization of urban areas. The restoration of brownfield sites will have to be part of a larger program of sustainable development and environmental stewardship in urban areas. Although brownfields in their degraded condition are in many ways environmental disaster areas, they also represent opportunities for significant, positive change in the local environment. Brownfield redevelopment has the potential to create jobs in economically depressed sections of urban areas both during and after the restoration of the area. Removing hazardous substances and decaying buildings will improve the environmental health and safety of brownfield sites, thus improving conditions for nearby neighborhoods. One of the most important benefits of brownfield restoration is the reuse of urban space. Such sites can be used for urban infill so that areas have new economic and human uses and thus can help slow the urban sprawl to exurban areas on the margins of our metropolitan developments. Along with economic redevelopment, restoring brownfield sites will allow us to restore green spaces to urban areas that provide ecological and social benefits to urban residents (Dorsey 2003).

Beyond the economic value of ecosystem services, there is also an increasing interest in ecosystem services as a conceptual tool that will allow us to more accurately model and hopefully better understand ecological restoration. Shahid Naeem (2006) thinks that examining biodiversity and ecosystem functioning will allow us to combine the strengths of community ecology and systems ecology approaches to ecological restoration. Community ecologists have traditionally looked at the species assemblages found in restored ecosystems and focused on biodiversity as the best measure of successful ecological restoration (as someone trained in community ecology, I can attest to taking that approach and knowing many others who do the same thing). Ecosystem ecologists have frequently been concerned with energy flows and nutrient cycles, but have not examined species composition in much detail. Combining the two approaches would provide a more complete picture of the ecology of restoration practice. Ecosystem services are not usually entirely contained within one ecosystem, and instead cross boundaries and occur at the landscape level. Examining ecosystem services at larger landscape scales will force restorationists to broaden their studies and to look for ways to better coordinate restoration projects located within a particular region in order to benefit from landscape-scale processes (Maurer 2006).

Restoration as a moral obligation

From the beginnings of the modern practice of ecological restoration, many restorationists have been motivated by what they see as a moral obligation to restore damaged environments. I discussed this obligation to a certain extent in Chapter 1, so I will not reiterate the points I made there, but I would like to emphasize a couple of key points.

In *The Sand County Almanac*, Aldo Leopold (1949) described his own evolution from happy wolf killer preoccupied with short-term gains in wildlife abundance to someone who began to take the long view and tried to think like a mountain in order to understand what was really best for entire ecosystems, not just humans. His work in the 1930s and 1940s was focused on developing a new moral relationship with the land in which we would treat the land as morally equivalent to ourselves and not just as a piece of property to be used and discarded.

Post-Leopold there has been increasing emphasis on the moral obligations we have toward the rest of the earth and much of the development of the entire field of environmental ethics has centered on determining what the exact nature of that obligation is. I do not have the space to make a lengthy examination of whether or not the natural world has intrinsic value. But for many restorationists who think that restoration is a moral obligation, whether the rest of the world has intrinsic value is not the source of the moral obligation. For them the moral obligation springs from a much simpler source, in many ways a return to the rules we learned as children – if we make a mess, then we are responsible for cleaning it up. We have a duty to set it right (Turner 1985; Janzen 1988). A key part of that duty is the fact that we are the only species that has the ability to set things right. In terms of environmental messes, unlike when we were children, there is no mother or father or teacher to step in and do the cleaning up for us if we fail in our responsibility. We are the clean-up crew. We could, of course, wait around and hope that nature will take its course and correct things – just as when we were children and the rain could be counted on to eventually wash away chalk drawings on the pavement. But we should also remember that many of our childhood messes were not so easily cleaned up. Our environmental messes are the same today.

We have the ability to do a lot of restoration, and as far as we know we are the only species with a moral code that would oblige us to do restoration, so the duty is ours (Turner 1985; Janzen 1988; Higgs 2003; Jordan 2003). It is possible to think of restoration as part of a developing sense of etiquette about how we should interact with the earth and its systems (Cheney and Weston 1999).

Eric Higgs (2005) worries that if ecological restoration becomes primarily focused on the technical aspects of restoration – if, say, all discussion of restoration is done in the context of more abstract scientific terms such as ecosystem functions and services – then the field will lose its respect for other sources of knowledge (especially social and cultural sources). Once social and cultural sources of knowledge are lost or at least are discounted, the science of ecological restoration will lose its moral center because science simply isn't meant to provide moral values.

I suspect that Higgs is right to be apprehensive about a potential loss of social, cultural and moral reasons for ecological restoration. That is one reason why a discussion of ecosystem values in primarily economic terms is troubling to me. There are some people for whom the need to do ecological restoration will be determined by direct human utility and the economic benefits it can provide. However, for many restorationists, ecological restoration is about something other than, and perhaps much greater than, utility. Remembering that for at least some of the pioneers of modern ecological restoration the practice of restoration was a moral obligation keeps the door open to thinking about restoration in broader terms. Even if the reason people do restoration starts with someone like Jay Stacy thinking "this is cool, I wonder how many different kinds of plants I can get growing here," or Rob Mungovan trying to restore the chalk stream of his childhood, those reasons spring from a particular moral view of the world. It may seem trite or cliché to say so, but sometimes the most important lessons we learn in life are the first ones we learn, and remembering we have a duty to clean up our messes is a powerful argument in favor of ecological restoration (Fulghum 1988).

4 Climate change

Is rapid pace and magnitude a bridge too far for ecological restoration?

In this chapter I will:

- briefly summarize the evidence for climate change;
- discuss model predictions of future climate change;
- examine ideas about the historical range of variability;
- examine chalk grasslands as a case study of restoration following changes to the historical range of variability.

The evidence for climate change

In Chapter 3 I claimed that the evidence for global climate change is overwhelming and that virtually all scientists are convinced global climate change is happening now and having major effects on our ecosystems (Millennium Ecosystem Assessment 2005; Parmesan 2006; Hansen *et al.* 2010; Lawler *et al.* 2010). So what are the effects of climate change that are most relevant to ecological restorationists? The most relevant effects are the ones that are most easily observed and that are likely to exert the largest changes on species distributions and behavior.

Changes in average temperatures lead directly to changes in species distributions as they move to areas that now have the correct conditions to support them, and are lost from areas that no longer have the proper conditions for survival, growth and reproduction. Every 1°C change in temperature causes ecological climate zones to shift by 160 km (in the northern hemisphere moving north with temperature increases) or to change in altitude by 160 m (moving up in altitude as temperature increases). There is already strong data demonstrating changes in species distributions around the world. In the northern hemisphere species have moved their northern distributional limits 6.1 km further north per decade over the past 50 years. Similarly, there has been an increase in plant species diversity in alpine regions of Switzerland as more plants move into higher altitude zones (Parmesan 2006; Thuller 2007).

Some of the most striking evidence that anthropogenic climate change is already leading to changes in species behavior comes from examining their phenology (initiation of growth, reproduction, senescence, etc.). A study of 1,598 species with good long-term records of phenology found that 41 percent exhibited changes in phenology consistent with global climate change (i.e., earlier spring flowering as spring weather becomes warmer earlier) (Parmesan 2006).

In the northern hemisphere species with seasonal activity have advanced their initiation of seasonal activity 2.3–5.1 days earlier per decade over the past 50 years (Thuller 2007). The two species with the longest detailed records of phenology – Japanese cherry trees (600 years) and European wine grapes (500 years) – both have significantly earlier flowering and harvest times than in the past (Parmesan 2006). Perhaps the most widely publicized data on changes in phenology are those relating to flowering time and initiation of seasonal growth among the plants in Thoreau's woods in Concord, Massachusetts. Henry David Thoreau carefully recorded the date of first flowering for over 500 species of plants from 1852 until 1858. At various times – 1878, 1888–1902, 1963–1993 and 2003–2006 – data for first flowering dates were recorded by other botanists and gardeners in the area. From 1852 until 2006, the Concord region has experienced an increase in average annual temperature of 2.4°C, and the plants are now flowering on average seven days earlier than during Thoreau's time (Miller-Rushing and Primack 2008).

What is especially interesting about the Thoreau's woods dataset is that the plant species that have experienced the greatest shifts in first flowering date have been the most successful at maintaining abundant populations. Those plant species that experienced the least change in first flowering date have experienced the most significant declines in population size since Thoreau's day. Plants that exhibited a lesser ability to shift first flowering date and thus suffered from the greatest population decline were mostly found within particular taxonomic groups – the anemones and buttercups (Ranuculaceae), asters and campanulas (Asterales), bluets (Rubiaceae), bladderworts (Lentibulariaceae), dogwoods (Cornaceae), lilies (Liliaceae), mints (Lamiaceae), orchids (Orchidaceae), roses (Rosaceae), saxifrages (Saxifragaceae) and violets (Violaceae) (Willis *et al.* 2008). It is especially troubling that some of the groups of plants which were not able to change first flowering date and which suffered population declines are groups of plants that are usually especially abundant and important members of many ecological communities, such as asters, mints, lilies, orchids and roses. If large numbers of those families decline, and even worse disappear, there would be serious changes in the structure and probably function of many of our ecosystems.

Interestingly, the owners of garden centers and nurseries have already adapted to these climate changes and successfully sell many native European garden plants far north of their original habitat – on average over 1,000 km farther north than their original home ecosystems (van der Veken *et al.* 2008). While gardeners carefully select varieties that will do well in more extreme conditions and also select sites in their gardens with ideal microclimates for their plants (i.e., planting a cold-sensitive plant in a sheltered spot near a wall that will collect solar radiation), the fact that native plants are successful in parts of Europe so far from their native ranges indicates that plants have the potential to adapt to climate change, but will almost certainly require assistance from humans to be able to move as rapidly as climate zones shift. It is estimated that plants in Europe will have to move 0.5–1 km per year over the next 100–200 years to track changes in climatic zones, but most plants do not disperse that far that fast on their own (van der Veken *et al.* 2008). Assisted migration – humans moving species to new areas in a race to beat predicted climate change – is a controversial topic that we will return to later, but it is something that is happening now in response to the human search for beauty and novelty in our gardens. There is also excellent evidence that global climate change will lead to changes in the distributions of invasive species which will make life more difficult for many native species already suffering reduced populations due to land-use changes (Pauchard *et al.* 2009; Kleinbauer *et al.* 2010).

Although the effects of climate change are easily observed in plant species, there is evidence of climate change affecting all taxonomic groups (Parmesan 2006). Many scientists and members of the general public have been especially concerned about declines to populations of amphibians, which are occurring all around the globe, such that about 32 percent of amphibian species are thought to be in danger of extinction (Millennium Ecosystem Assessment 2005). There are many reasons for the decline of amphibians, including habitat loss and the spread of especially injurious diseases among amphibians, but there is also direct evidence of amphibian declines due to climate change. In Yellowstone National Park, the world's first national park and an area that is usually thought to be especially well protected, increasing temperatures during the past 100 years have led to many ponds either decreasing in size or drying up completely. The declines in pond size and number are significantly correlated with declines in both the number of amphibian species and individuals in Yellowstone National Park (McMenamin *et al.* 2008).

There has also been a great amount of concern about the effects of global climate change on pollinator species and the plants they pollinate. We are already seeing changes in pollinator behavior due to climate change in temperate zones. On average, the date of first flight activity for many pollinating insect species occurs four days earlier per 1°C increase in temperature (Memmott *et al.* 2007). This advance in pollinator activity correlates well with observed advances in flowering time for many temperate-zone plants (Memmott *et al.* 2007; Thuller 2007; Miller-Rushing and Primack 2008). The fact that at the current time pollinators and plants are exhibiting changes in the initiation of activity which are moving in the same direction should not lull us into a false sense of security that pollinators and plants will adapt similarly to global climate change. Jane Memmott and colleagues conducted a sophisticated modeling study that examined how projected climate change due to a doubling of atmospheric CO_2 would affect a temperate-zone network of 1,420 pollinators and 429 plant species. They ran several simulations and there was some variability in the overall results, but the trend was the same in all simulations – there are predicted to be declines in 17–50 percent of floral resources available to pollinators and as many as half of the dates when pollinators were predicted to be active would occur when none of the plant species they are known to use would be flowering (Memmott *et al.* 2007). They predicted that the result would be extinction of the affected plants, pollinators and their interactions – a situation that in 1974 Dan Janzen said would result in an especially egregious kind of extinction – the extinction of ecological interactions (Janzen 1974). Although Memmott and colleagues were careful to say that their modeling results applied only to the pollinator–plant network that they modeled (a network that occurs in the prairie–forest transition in western Illinois) and that the modeling results are not 100 percent certain to occur, it is not difficult to imagine or hypothesize that similar losses in plants and pollinators are likely to occur in all ecosystems affected by climate change.

Marine systems have also been affected by current climate change. Average sea level has risen about 1.8 mm/yr over the past 100 years, and on-going climate change is expected to result in 8–88 cm of additional sea level rise over the next 100 years. Many coastal ecosystems and their constituent species, such as coastal salt marshes and mangroves, are already suffering stress and reductions due to habitat destruction and pollution. An increase in global sea levels will exacerbate their decline because coastal areas around the world are highly developed, home to most of the human population and thus there may be no open space for coastal ecosystems to retreat to as sea level rises. During the last 50 years about 20 percent of coral reefs have been destroyed and a further 20 percent degraded. Much of this loss of coral reefs is due to siltation, pollutants, habitat destruction and direct exploitation, but there have been losses due to climate change. Many areas of coral reef have suffered from bleaching events in which sea water temperatures increased by 0.5–1°C more than the average temperature of the hottest month (Millennium Ecosystem Assessment 2005).

Thuller (2007) claimed that understanding the process of climate change, its consequences for species and ecosystems, and how we should respond to climate change is the "Grand Challenge" for ecologists in the twenty-first century. In 1990 Daniel Botkin warned us that in the twenty-first century the natural world will be a nature that we humans have made. The question we must address is whether we will make that nature intentionally or unintentionally and whether that nature will be desirable or undesirable (Botkin 1990). If we are to intentionally create a desirable world, then ecological restoration will have to be one of the tools we use to build the twenty-first century. But as we will see in the following sections of this chapter, there are many uncertainties that must be addressed before we have any hope of using ecological restoration to mitigate climate change.

Model predictions of future climate change

We use current evidence of global climate change along with models of the potential for future change to attempt to understand how climate change will affect species and ecosystems during the next 100 years and beyond. The models that have been developed all produce results that are in general agreement about large-scale changes in future climates – additional increases of greenhouse gases, especially CO_2, will result in the average global temperature increasing by 2–6.4°C above the average in 1750 by 2100 (Millennium Ecosystem Assessment 2005; Saxon *et al.* 2005; Thuller 2007; Hansen *et al.* 2010; Kiesecker *et al.* 2010, Lawler *et al.* 2010). However, it is extremely difficult to predict what changes will occur at a specific location, especially when we add in other factors along with changes in temperature. Organisms live in environments in which many physical factors influence their survival, growth and reproduction, and for terrestrial organisms the most important physical factors after temperature are precipitation and relative humidity. Model predictions are tremendously variable regarding how precipitation patterns will change at both broad continental and smaller regional scales. This variability in predictions about precipitation is at least partly due to the many different models in use to predict the effects of climate change. At the current time, there are at least 24 different atmosphere–ocean general circulation models used to predict future climate changes. The different models make slightly different assumptions about the direction and rates of future changes and employ different algorithms to make their calculations. As a result, we can find models that predict precipitation will increase in a specific region and others that predict precipitation will decrease for the same region (Lawler *et al.* 2010).

We have already observed increases in global temperature that are occurring more rapidly than predicted in 2007 by the IPCC (Intergovernmental Panel on Climate Change) models (Hansen *et al.* 2010). Current models predict that in order to keep the global average temperature from increasing 2°C (the lowest expected increase in almost all model results) we need to ensure that carbon emissions reach their peak during the 2010s and are reduced to 80 percent of the 1990 levels by 2050. Almost all experts agree that we are highly unlikely to achieve those carbon emission goals (Galatowitsch 2009). Thus we can almost certainly expect to see average global temperatures climb by more than 2°C in the next 100 years. Given that depressing scenario there is broad agreement among modelers that over the next 100 years, the climate will change, temperatures will increase and precipitation will almost certainly become more variable, at least in the short term as the global climate sorts itself out as a result of internal and external changes and feedback loops (Millennium Ecosystem Assessment 2005; Saxon *et al.* 2005; Thuller 2007; Hansen *et al.* 2010; Kiesecker *et al.* 2010; Lawler *et al.* 2010).

Ecologists developed many models to predict changes in ecosystems throughout the twentieth century, and during that time the models became increasingly sophisticated. However, for most of that time the models were based on the assumption that ecosystems were basically stable – that if perturbed they would return to their original starting condition (Holling 1979; Hansen *et al.* 2010; Lawler *et al.* 2010). Even today many of our current management plans for nature reserves all across the earth are based on rather old ideas of static ecosystems that should always tend to return to the original condition following perturbation (Hansen *et al.* 2010; Lawler *et al.* 2010). Probably the greatest challenge for ecosystem modelers today, and the main reason the results of different models may be at odds with each other, is the struggle to develop models that incorporate constant change as a part of the new norm for ecosystem functioning. However, the current model predictions are in enough agreement about general trends and the predicted changes are extreme enough that they are more than a little frightening, both in terms of how the changes will affect our personal lives and in terms of how the changes will make our task as ecological restorationists drastically more difficult.

Because of personal and professional interest my starting point in trying to understand predictions of climate change is to first examine the predicted impacts to my home region in Illinois. The Union of Concerned Scientists has put together a series of graphic images of climate migration that show how predicted changes in climate will result in the future Illinois having a climate that is similar to the climate that currently exists in more southern regions of the United States (Figure 4.1).

Their prediction is that by 2095 Illinois will have a summer climate that is similar to that in east Texas today – hotter and drier than today in Illinois – and also that by 2095 Illinois will have a winter climate that is similar to western Arkansas, eastern Oklahoma's present winter climate – a warmer more humid climate than we currently experience in the winter. I personally do not like very hot summer weather and for me that is enough reason not to live in east Texas. But I won't be around in 2095, so perhaps I shouldn't worry about that predicted change – it won't affect me. But there is a chance my children will be here then, and if they have children, it is highly likely that my grandchildren will be alive in 2095. If only for the sake of my descendents I have to be concerned about those future climates. Not only will the summers be less pleasant, but there will almost certainly be increases in energy consumption or at least demand for energy (assuming there is still access to relatively affordable energy) as

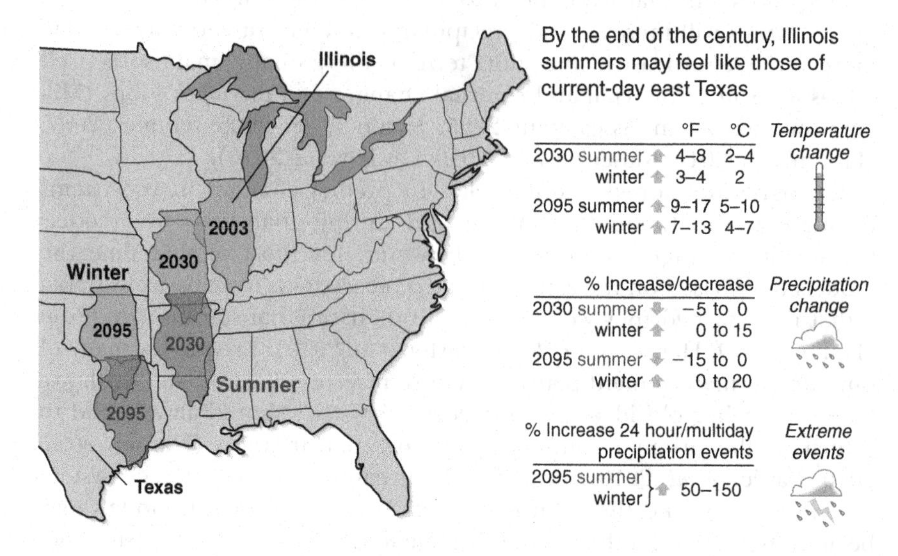

Figure 4.1 Map of the central and eastern portions of the United States showing how the climate for the state of Illinois is projected to change between 2000 and 2095. Figure used by permission of the Union of Concerned Scientists (source: www.ucsusa.org/greatlakes/glimpactmigrating.html accessed November 17, 2010).

Note

The map shows the state "migrating" from its current climatic position to the regions of the United States that today have a climate similar to what Illinois will have in 2095.

people attempt to cool their houses. There will be changes in agricultural practices as the crops grown in Illinois today – exceedingly important on both a national and global level – may not do well in the predicted Illinois of 2095.

But as frightening as the predictions from the Union of Concerned Scientists are, there are other models that predict a future Illinois with even more extreme climate changes. Saxon and colleagues (2005) prepared models that examined changes in ecological domains for the continental United States and part of lower Canada, and also examined some more local areas in greater detail. One of the areas they explored in detail was the Midwestern corn (maize) belt – which is centered on Illinois. They ran simulations at two levels of increased atmospheric CO_2 – moderate emissions with CO_2 reaching 545–770 ppm by 2100 and high emissions with CO_2 reaching 735–1,080 ppm by 2100. Under the moderate scenario Illinois is predicted to have a mean temperate of –3.3°C for the coldest quarter (vs. –8.9°C now) and a mean temperature of 41.4°C for the hottest quarter (vs. 29.3°C now). Under the more extreme high emissions scenario the coldest quarter is predicted to have a mean temperature of 1.7°C and the hottest quarter a mean temperature of 48.4°C (Saxon *et al.* 2005).

I was discussing the model predictions with Jim Harris, and when I mentioned a mean summer temperature of 48.4°C, he burst out, "But that's like Iraq!" And that really brought home the potential for change to local climates – Illinois transformed to a summer temperature regime like the current summers in Iraq is an Illinois that I would find a difficult place to live. Even though it is likely that Illinois would have more precipitation than does Iraq (but less summer precipitation than Illinois currently receives), it is unlikely that our current varieties of maize and soybeans – so important to agriculture in the corn belt – would prosper there either. As I think about the tallgrass prairie restorations I manage in western Illinois, I can't imagine them thriving in either the Union of Concerned Scientists' prediction of Illinois becoming east Texas or Saxon and colleagues' prediction of Illinois becoming Iraq. Those restorations would not be viable under those scenarios and that knowledge forces me to drastically reconsider my goals and responsibilities with respect to those restorations.

Saxon and colleagues' (2005) examination of climate change in North America is fascinating in a slightly sickening, gut-churning way for both North Americans and others from elsewhere around the world. They predicted that under the moderate change scenario 13.3 percent of current "ecological domains" (their term – essentially local eco-climatic regions) would be lost and 53.6 percent of the modeled area in North America would have new ecological domains. Under the more extreme high emissions scenario, 63.2 percent of current ecological domains would be lost and 63.1 percent of the modeled area would have new ecological domains. The new ecological domains are especially intriguing/troubling (I suppose your view will depend on whether you are a glass-half-full or glass-half-empty person) because the new ecological domains have no current analogue in North America – i.e., they represent combinations of temperature, precipitation and humidity that do not currently exist there. As we shall see, no-analogue ecosystems developing during predicted climate change are a common result of models examining changes to ecosystems at the local and global scale. These no-analogue conditions represent huge challenges to restorationists because they may render unviable many current restoration projects and force us to revisit our goals and aspirations for ecological restoration in the twenty-first century.

I live and work in western Illinois, but whenever possible I travel to the north woods around Lake Superior. I especially like to go to those forests during August so I can escape the dog days of summer in Illinois. I also like to go there in mid-winter so I can walk or ski through forests that lay silent under a deep blanket of snow. Because this is another ecological region that I know well, examining model predictions for the north woods also helps me visualize how global climate change will affect regional ecosystems.

Ravenscroft and colleagues (2010) developed models that allowed them to examine possible changes to the forests that occur in northeastern Minnesota – a region that lies along the north shore of Lake Superior and extends inland to encompass the entire Boundary Waters Canoe Area and parts of the old Iron Range. Forests in this region are transitional, occurring along an ecotone that marks a change from temperate deciduous forests (dominated by maples, birch and basswood along with some black ash and a few oaks and hickories) to truly boreal coniferous forests (dominated by red, white and jack pines, white and black spruces, and balsam fir, depending on soil type and topography). Their models predict that there will be large changes to the composition of forests in northeastern Minnesota, with an almost complete loss of white spruce, balsam fir and paper birch, regardless of whether the models used moderate or high levels of CO_2 emissions. Under high CO_2 emissions there would also be a loss of almost all black spruce, red pine and jack pine (Ravenscroft *et al.* 2010). Their models predict that there would be an expansion of both sugar and red maples across the region and that where today the forests exist in a mosaic of various forest types, in the future the forests would be a much more homogenous stand of maples. Maples are very beautiful trees – especially in the autumn when dressed in their crimson finery and an expansion of maples would probably be a boon to the syrup industry – and perhaps a boon to all of us if maple syrup production increased enough to drive down the prices of maple syrup. But when I think of the north woods and what I value most about the forests there, I can't imagine the north woods and the Superior lakeshore without pines and firs. I, and I suspect most people in the region, love to walk beneath the pines on a soft bed of pine needles and to smell the rich incense when sitting or sleeping below a balsam fir. The loss of those species and experiences would create a vastly different sense of what the Lake Superior region means to me and many others. It might not necessarily be worse, but it would definitely be different.

Ravenscroft and colleagues (2010) were especially interested in determining whether particular management schemes could preserve the north woods in the future. Their main conclusion was that, given the predicted levels of climate change, it will be impossible to manage the north woods in Minnesota in a way that would maintain the historical range of variability (HRV). White spruce, balsam fir and paper birch would be lost regardless of the management scheme applied. Interestingly, their models predicted that even without any climate change, the current management practices are inadequate to preserve the full range of diversity of the north woods. Clearly we have a lot to learn about managing our woods even without the specter of climate change looming over our heads.

In order to appreciate the impact of environmental change on a global scale, it is worth considering what kinds of environmental change have occurred in the recent past. The Millennium Ecosystem Assessment (2005) provided an excellent summary of recent changes to the earth's ecosystems. The overall picture painted by the Millennium Ecosystem Assessment team was far from rosy. The period from 1950 until 2000 saw unprecedented environmental change. The changes were driven by increasing human population – which doubled from three billion to six billion during that period. The rapid increase in our population resulted in ever-increasing pressures to provide food, shelter, clothing and other basic needs to all of us. Supplying those needs led to a huge increase in development of the earth's surface and the use of fossil fuels and various chemicals. The period from 1950 until 1980 saw more land converted to agricultural production than the period from 1700 to 1850, when the introduction of new crops to Europe (especially potato and maize) and the beginning of industrialization also saw a rapid increase in human population and expansion of agriculture. Today, about 25 percent of the earth's land surface is used in food production in some way or another. The Industrial Revolution was largely powered by increased burning of fossil fuels, which resulted in an increase of atmospheric CO_2 from 280 ppm to 376 ppm by 2003. About 60 percent of that increase occurred after 1959, and the rate of use of fossil fuels and release of CO_2 into the atmosphere continues to increase today.

The need to produce increasing amounts of food has led to the use of more and more chemical fertilizers, especially nitrogen and phosphorus, much of which has found its way inadvertently into domesticated and wild ecosystems, resulting in pollution by excess nutrient input. Since 1960 the amount of biologically active nitrogen used annually has doubled and phosphorus has tripled. The Millennium Ecosystem Assessment team found that 60 percent of the ecosystem services they measured have declined in the past 50 years as a result of these and other changes to the environment. The primary driver of environmental change has been changes in land use and habitat modification, but in the twenty-first century global climate change will have a larger and larger role to play in driving environmental change.

Given that the human population is expected to reach nine billion by 2050 and that we are failing to reduce our use of fossil fuels, we can expect the rate of environmental change to accelerate. Although the Millennium Ecosystem Assessment team noted that environmental change is usually assumed to be gradual, they acknowledged that in the future there is the potential for environmental change to become more sudden, proceeding by abrupt non-linear shifts in conditions from one state to another. In fact, they report that there is some evidence that such abrupt shifts are already occurring, as seen in sudden collapses of fisheries like the Newfoundland cod fishery, drastic changes to the Great Lakes ecosystem caused by the invasion of zebra mussels and feedbacks between deforestation and declines in precipitation. Deforestation often results in a decline in local precipitation because of a large decrease in evapotranspiration and thus a loss of moisture returning to the atmosphere in the deforested region. As deforested areas get larger, the region suffering a decline in precipitation gets larger and larger in size and the lack of precipitation makes it more difficult, if not impossible, for tree seedlings to survive. So we observe that there is no forest recruitment, thus continued increase in evapotranspiration and less precipitation, etc., until the whole region may be transformed from a forest to dry grassland.

All of the climate modelers agree that the potential changes to the global climate are difficult to predict with a high degree of certainty. But what is certain is that we will have to be both flexible and proactive in our management of domestic and natural resources as we adapt our lives and livelihoods to the new environmental norm that will develop over the next 100 years (Millennium Ecosystem Assessment 2005; Thuller 2007; Hansen *et al.* 2010; Kiesecker *et al.* 2010; Lawler *et al.* 2010).

Caveats and challenges when using models to predict global change

As noted above, the various models used to predict future climate change are based on a range of assumptions and thus produce results that sometimes disagree with each other. Good modelers are explicit about the uncertainties built into the models they develop and the predictions that result from their models (Millennium Ecosystem Assessment 2005; Saxon *et al.* 2005; Hansen *et al.* 2010; Kiesecker *et al.* 2010; Lawler *et al.* 2010). Critics of the research into global climate change have seized on disagreement about specific details among different models and the fact that models are probabilistic and thus always incorporate uncertainties. This has been especially true in the United States, where there is a vocal political movement composed of people who disagree with the very idea that human actions can cause climate change. What climate change deniers have failed to acknowledge is that despite disagreements about specific details, the models are in broad agreement about the general trend of increasing average global temperature and large changes to global, regional and local patterns of precipitation and seasonal variation. However, scientists and policy-makers studying global environmental change would do well themselves to acknowledge that some of the uncertainties and variability in model results is also troubling to them. In fact, some of the harshest critics of the current generation of climate change models are scientists who are convinced that climate change is happening and that major, negative effects of climate change are almost certainly unavoidable.

James Lovelock, in particular, thinks that the IPCC model predictions of climate change during the next 100 years are of little use because those models are based on a smooth trend of gradual warming (Watson 2009). Lovelock has long expounded on his view that the earth's climate is intimately connected to the biology of the land and sea. Because the ecology of land and sea are experiencing rapid change due to habitat loss and destruction, pollutants, invasive species and the effects of on-going climate change, he thinks that real climate change is unlikely to be smooth. Instead, he predicts that real climate change in the next 100 years is likely to occur by sudden abrupt shifts in which climate flips from one state to another once critical thresholds or tipping points are reached. Once we cross these tipping points, the new environmental norms are likely to be irreversible in terms of many, many human generations. Although the Millennium Ecosystem Assessment (2005) mentioned the potential for abrupt non-linear changes in environmental conditions, this is not an area that has been widely explored by climate modelers. In the context of social systems we have learned that the great problem with tipping points is that we usually don't recognize them until they have been passed, the entire system has shifted and frequently it is too late to do anything to prevent or correct them (Gladwell 2000). Lovelock, who has become increasingly pessimistic as he has aged, thinks we are rapidly approaching a future in which there will be huge migrations of people in search of better conditions and massive human mortality after the climate crosses a tipping point (Watson 2009).

In many ways apocalyptic predictions of future conditions feed into the denials issued by climate change skeptics. They claim that we have seen many predictions of environmental collapse and famine in the past and these predictions have not come to pass – so why should the new predictions be any more believable? They may have a point. An alternative future is one in which drastic climate change coupled with increasing human population forces us to convert every bit of land surface to either living space, agricultural land or industrial centers (Watson 2009).

I have heard Dan Janzen give talks in which he predicts that the future earth will look like a giant version of Iowa. And despite my fondness for Iowa (I met my wife there and it is a particularly pleasant place), I have to acknowledge that Janzen is not being complimentary when he makes that statement. For Janzen, Iowa is a land that is completely domesticated – he sees it as a place where every inch of space is either used in agriculture (primarily to grow maize and soybeans) or urbanized. My own sense is that a future, completely domesticated earth is a more likely scenario than an apocalyptic earth – one thing we humans are good at is survival. And if survival means domesticating the entire earth, we will do so. The end result will be a boring earth with no wild spaces and very little biodiversity – an impoverished earth in which our vast, beautiful biological heritage has been squandered. It would be an earth even more in need of ecological restoration than our current earth – but what would we have left to work with? With the majority of species extinct, open, unused space non-existent, there would be no scope for restoration work. It is a depressing thought.

So our models aren't perfect. So what? Our best response to deniers of climate change is to explain that our current models miss the potential for tipping points and thus are too optimistic. If anything, the reality of climate change may be more extreme than our models currently predict.

Historical range of variability: a now unattainable goal?

Ecosystem managers have been concerned about the range of natural variation at least since the 1960s (Landres *et al.* 1999). Their concern arose from questions as to whether the changes they observed in ecosystems – both in terms of composition and function – were similar to past changes that would be expected due to normal levels of variation within a dynamic system, or whether the observed changes were so large as to be outside the range of expected variation. Because our traditional approach to managing ecosystems has been to maintain them in their current condition, we wanted to be able to recognize when changes exceeded the range of variation experienced in the past and thus represented significant departures from what might be considered normal ecosystem variation. Our interest in determining the historical range of ecosystem variation stemmed from a worldview that natural, healthy ecosystems should be at something approaching equilibrium. We thought that any changes to ecosystem composition or function must exist within a fairly well-defined range of expected variation if the ecosystem was to continue to exist at equilibrium conditions.

We usually manage natural systems in order to satisfy many human needs and desires for such things as flood control, nutrient cycling, a continuous supply of natural products, recreational opportunities and even the maintenance of the ecosystem in a natural condition. Conservationists and restorationists have traditionally been interested in managing our ecosystems in order to preserve as much naturalness of ecosystem composition and function as possible. Managing for naturalness seems like an oxymoron, but it reflects the fact that we try to preserve ecosystems as close to their historical natural condition as possible, while also preserving features and services we think desirable. In the designated National Wilderness Areas in the United States, this led to a management program of attempting to maintain the wilderness areas within boundaries known as "limits of acceptable change" (Brunson 2000). The framework for the limits of acceptable change management approach was also based on the idea of an equilibrium ecosystem such that even though ecosystems are dynamic and changeable, they cannot change too much and remain the same ecosystem. At some point, extreme change must lead to the formation of a new type of ecosystem. The major, in many ways the only, goal of the US Wilderness Protection Act, which established the National Wilderness Areas, was to preserve large tracts of land in as pristine a condition as possible and as close to historical conditions (the conditions that existed at the time of Euro-American settlement) as possible. However, the concept of limits of acceptable change would be familiar to any ecosystem manager attempting to maintain a specific site within a defined historical range of variation. The management of ecosystems based on maintaining the ecosystem within the historical range of variation often requires ecological restoration in order to maintain the historical conditions. As the ecosystem is observed to approach or exceed the historical or accepted range of variation, the manager has to take steps to ensure the ecosystem remains within or returns to the desired range, thus necessitating restoration.

If we employ a concept like the historical range of variation when managing an ecosystem, we are forced to characterize the ecosystem as existing in a particular set of spatial and temporal scales with respect to variation in composition and function. In the past, human activities such as grazing animals on ecosystems like chalk grasslands in England helped to maintain the chalk grassland ecosystem within a particular range of variation – a range of variation that has existed for thousands of years (Rackham 1986). In the last few decades human activities both within ecosystems (changes in land use) and external to the ecosystem (burning fossil fuels releasing CO_2 so that the entire atmosphere and climate are changing) are increasingly pushing ecosystems outside their historical range of variation (see the case study on chalk grasslands, below). In fact, environmental changes are so drastic that virtually all ecosystems have been impacted and many, if not most, now exist in conditions that are beyond their historical range of variation (Millennium Ecosystem Assessment 2005; Seastedt *et al.* 2008). These changes have created a situation in which most ecosystems are hybrid or novel ecosystems or will soon become hybrid or novel ecosystems (novel ecosystems will be discussed in detail in Chapter 5) (Seastedt *et al.* 2008; Hobbs *et al.* 2009). Our old models of equilibrium ecosystems and management goals of maintaining ecosystems within limits of acceptable change and historical ranges of variation are becoming less and less applicable in the twenty-first century, and soon will not be applicable at all. Restoring ecosystems to the historical ranges of variation is becoming an increasingly unattainable goal. From now on, management and restoration of novel and hybrid ecosystems essentially will be experiments, and scientists, managers and stakeholders must collaborate to ensure that we develop management plans and goals that are attainable and that meet the needs of humans and all of the biota.

David Lowenthal neatly summed up our management dilemma in his 1985 book, *The Past is a Foreign Country*:

> Recognising the impact of the present on the past, we confront anew the paradox implicit in preservation. Vestiges are saved to stave off decay, destruction and replacement and to keep an unspoiled heritage. Yet preservation itself reveals that permanence is an illusion. The

> more we save, the more aware we become that such remains are continually altered and re-interpreted. We suspend their erosion only to transform them in other ways. And saviours of the past change it no less than iconoclasts bent on its destruction.

Any management action we take will result in changes to the ecosystem – even if we are trying not to change it all. Moving beyond historical ranges of variation as a primary management goal is not abandoning our past traditions of management as much as it is recognizing that regardless of what we do, the system will change. Thus our planning needs to be focused on the future so that hopefully the changes that do happen are changes that are desirable (or the least undesirable in a worst case scenario) and that the ecosystem continues to satisfy our needs and the needs of all living things.

Case study: restoration and management of chalk grasslands

Chalk grasslands are semi-natural grasslands found on shallow, calcareous, low-nutrient soils located across northern, especially northwestern, Europe. Chalk grasslands are a human-created ecosystem. They occur in areas that were originally forested and that were cleared by humans, in some cases as long ago as the Bronze Age. They evolved under a system of low-intensity agriculture characterized by cattle and sheep grazing. In some areas there was also occasional mowing for hay production or the occasional use of fire to prevent the incursion of woody plants. They have been maintained for centuries, even thousands of years, by continuous low-intensity agriculture, especially continuous grazing. As a result of their development under a regime of human use, their continued existence is completely dependent upon a continuation of traditional agricultural practices. The long period of evolutionary development and continued existence of chalk grasslands created a unique ecosystem that supports a highly diverse characteristic flora and fauna in which many of the species are seldom found in other ecosystems. Some authorities claim that chalk grasslands are the most species-rich ecosystems for plants and invertebrates on earth when measured at a small scale – say $10\,m^2$ or less. However, despite their unique and diverse species assemblages, chalk grasslands have been in decline ever since the advent of high-intensity agriculture at the end of World War II (Thompson *et al.* 1999; Willems 2001; Woodcock *et al.* 2005).

Prior to World War II, chalk grasslands were farmed by traditional low-intensity grazing for centuries to thousands of years because the soil was relatively shallow, rocky and low in nutrients. Other types of agricultural use simply did not make any sense in such poor ground. Low-intensity grazing could be used to reliably supply meat, milk, wool and dung, which could be used for both fuel and fertilizer on more productive land. However, after the war the development of relatively inexpensive chemical fertilizers led to drastic changes in farming practices across Europe. Chalk grasslands suffered three different kinds of changes to their traditional use and management – all of which pushed them outside their historical range of variation. In many cases chalk grasslands were plowed, fertilizers were applied and arable row-crop agriculture was initiated. In other cases fertilizers were directly applied to the grasslands, resulting in increased productivity of the plants and richer, taller growth of the grasses. Increased plant productivity allowed both higher stocking rates of sheep and cattle and also the growth of the grasses to produce more abundant hay crops. However, the increased productivity resulted in a decline of species richness as a few grasses like erect brome (*Bromus erectus*) came to dominate, and species less able to take advantage of the nutrient input were out-competed and disappeared. Finally some chalk grasslands were simply abandoned and no longer grazed or mown for hay. Abandonment happened most often to steep or hilly sites that were difficult to farm using machinery, or small plots that were not economical to farm with intensive methods. Because abandoned grasslands were no longer grazed they often became dominated by woody vegetation. In Britain the abandonment of grasslands was exacerbated by the outbreak of myxomatosis in the 1950s, which killed most of the rabbit population. Thus, after sheep and cattle grazing ended, there was no other grazer available to replace them, which allowed a rapid expansion of woody vegetation. Because of the loss of chalk grassland and the ensuing decline of their characteristic species, chalk grasslands were listed as habitats of special concern in the Habitats Directive (EC Council Directive on Conservation of Natural Habitats and of Wild Fauna and Flora (92/43/ECC)). Since then there have been many efforts to preserve existing chalk grassland and to restore damaged or degraded sites (Thompson *et al.* 1999; Willems 2001; Woodcock *et al.* 2005).

Efforts to restored degraded chalk grassland vary depending on the source of degradation (Thompson *et al.* 1999; Willems 2001; Woodcock *et al.* 2005). Abandoned sites are cleared of encroaching woody plants. Areas converted to arable agriculture are usually smoothed and then either planted with native chalk grassland seeds or allowed to seed naturally. Sites that received increased fertilizer inputs are particularly difficult to restore. The effects of excess nitrogen fertilizers decline within 10–15 years of fertilization ending and a resumption of grazing and/or mowing. Most of the excess nitrogen is removed from the system by plant uptake and regular grazing and mowing to remove their growth. However, the effects of excess phosphorous fertilizers are more long-lasting, persisting for at least 25 years after the end of fertilizer application, despite various grazing and mowing regimes. In severe cases the highly fertilized topsoil can be removed, but that is expensive and destroys the remaining plant community, so it is not often recommended. Unfortunately, the continued aerial deposition of atmospheric nitrogen results in unintended fertilizer input that causes continued excess productivity and thus continued decline of some species due to being out-competed by faster-growing plants (Smits *et al.* 2008).

I have read about chalk grasslands, their amazing biodiversity and their restoration for years and was eager to visit a chalk grassland restoration site. My first opportunity to do so came on a trip to the Pegsdon Hills Nature Reserve near Barton-le-Clay, Bedfordshire, UK. I was fortunate to have two excellent guides with me that day – Graham Bellamy from the Wildlife Trusts and Heather Webb from the Bedfordshire County Council. I was fortunate for two reasons – first to learn the history of the restoration firsthand from Graham, who has been involved with the project since its inception; and second because without them I would not have known more than a handful of the plant species there. As a plant ecologist it is always humbling to go to a new ecosystem in a new country and to look at the plants and think, okay, they all look vaguely familiar, but I can't identify any of them any closer than the family. This was especially true because our visit was on a chilly November afternoon, when the only plant in flower was autumn gentian (*Gentianella amarelle*) and all the plants had been grazed fairly close to the ground. I kneeled on the cold, damp ground, my knees becoming coated in an unfamiliar gray, chalky mud and thought to myself, "This is obviously very diverse – I'm seeing a huge variety of leaf shape, texture and color, but I don't know any of these plants." Humbling yes, but it also provided inspiration to learn more – and luckily my guides were there to teach me.

Pegsdon Hills (Figure 4.2) is a classic example of the development and exploitation of chalk grassland. The site is almost 200 acres and occurs along a steep vertical gradient from flatland in the valley bottom along the B655 to the top of the Chiltern Hills. It has been grazed since the Bronze Age, at first mostly by cattle, then for the last couple of hundred years mostly by sheep. During World War II, the lower, flatter areas (about two-thirds of the

Figure 4.2 A view across the Wildlife Trust's Pegsdon Hills chalk grassland restoration.

Note
This view is from about halfway up the hillside, looking across a 4.5 ha experimental site in which different treatments are enclosed in fencing. The restoration site extends down the hill to a line of trees and buildings (seen in the middle of the frame), which occur along the B655.

reserve) were plowed to produce food needed during the war. Afterwards, arable crop production continued until 1992, with the last few years of arable production being focused on growing rapeseed for oil production.

The steeper areas have probably never been plowed, but suffered from woody encroachment as grazing declined following the war. Restoration efforts began in 1992. Some of the formerly plowed areas were planted with chalk grassland seed, other areas were allowed to seed naturally from nearby chalk grasslands. There has also been some recovery via the seed bank – especially by cowslip *Primula veris.* Today the site is managed mainly by grazing. The Wildlife Trust uses a combination of cattle, shorthorns and Belted Galloways, and Hebridean sheep (Figure 4.3) and a Manx sheep that both do well with poor forage. Different species and even breeds of grazers have differing effects on the plant community because they preferentially select different species and graze closer or farther from the soil surface, so using varying grazers or several types enhances heterogeneity and species diversity (Woodcock *et al.* 2005). Ironically, there is a small, two-acre site at Pegsdon in which the soil is regularly disturbed to mimic plowing in order to maintain several species of arable field weeds that are now becoming rare due to the use of herbicides.

Pegsdon Hills is an impressive site. Partly that is because of its spectacular natural beauty, especially if you hike to the top of the Chilterns and gaze down toward the valley floor. As we stood on top of the hills, we watched red kites (*Milvus milvus*) soaring below us. And partly it is because of the variety of species that occur there. Even though I couldn't name the many species present, it was obvious that Pegsdon is very species-rich. Interestingly, there doesn't seem to be any difference in plant species richness between sites that

Figure 4.3 Some of the Hebridean sheep at Pegsdon Hills, which are managed to create the traditional grazing practices that maintained the chalk grassland for hundreds of years.

were planted with seeds and sites that seeded naturally – they seem to have matched each other. Pegsdon also supports many interesting invertebrates, in particular yellow meadow ant (*Lasius flavus*) which is extremely abundant there. In unplowed sections the ants have made huge colony mounds – many 40–50 cm wide and 20–30 cm tall (Figure 4.4). In some places they are as abundant as one colony per square meter. They must have a huge influence on below-ground soil processes. Graham thinks the largest colonies must be decades old – perhaps 60–100 years. Colonies are starting to develop in formerly plowed areas. Of course, Pegsdon is like all restoration projects; there are always small problems arising. Brambles are starting to become established on some steep areas and need to be cleared by hand. There has been a recent outbreak of rabbits, and their grazing has denuded some areas. And even though Pegsdon has high species diversity, it still doesn't match nearby, undisturbed reference sites like those at the Knocking Hoe Nature Reserve. Heather wonders how long complete recovery will take; in a paper posted on her blog she estimates that it may take at least 100 years to fully recover biodiversity and community structure, if full recovery is even possible (Webb 2010).

As we walked down the steep Chilterns, our boots and trousers coated with chalky dust, chilled from standing out in the cold wind whipping along the ridge top, what I found most thought-provoking about Pegsdon (and many restoration projects) is that a few years of human use (here about 52 years of arable agriculture) resulted in damage that will take far, far longer

Figure 4.4 A large mound marking the location of a colony of the yellow meadow ant (*Lasius flavus*).

Note
This large mound is typical of the many abundant ant mounds found in unplowed sections of Pegsdon Hills. The mounds create micro-habitats colonized by a variety of plants and the ants themselves are important in below-ground ecosystem processes.

to correct. Pegsdon is now being managed with the same low-intensity grazing it received for hundreds, probably thousands of years, but it will be a long time before it is back to the ecosystem shaped by that earlier management. If only it was as easy to restore habitat as it is to destroy it.

That brings us back to coming climate change. Given the long time necessary to fully restore chalk grasslands and the rapid pace of climate change, is full restoration even possible? Many chalk grassland species are well adapted to the current, somewhat cool, moist climate. How will they fare as Europe becomes hotter and drier? And will it be possible to consistently manage chalk grasslands in a way that maintains their current condition while also allowing them to evolve with coming climate change? Pegsdon Hills is a unique site because it is a large nature reserve. But most chalk grassland sites are smaller and many are privately owned and maintained by farmers enrolled in agri-environment agreements (Thompson *et al.* 1999). Will coming climate change create pressures on farmers that will make it difficult for them to continue the traditional low-intensity grazing necessary to maintain chalk grassland? Once again, our knowledge of coming climate change leads to more questions than we can currently answer without a crystal ball.

5 Novel ecosystems

A new wrinkle for ecological restoration

In this chapter I will:

- describe the development of novel ecosystems and their relationship to climate change;
- relate ecological restoration to the tradition of adaptive management;
- discuss managed relocation as a special type of restoration;
- describe unique opportunities and special challenges that will arise for restorationists in the twenty-first century;
- conclude by looking at renewed restoration as a new path in the twenty-first century.

Novel ecosystems and climate change

Climate change and the accompanying changes to ecosystems are nothing new for our earth (Botkin 1990). In fact, climate change is the norm for the earth's past, particularly during the Pleistocene when repeated glacial cycles led to drastic changes in the global climate over fairly regular intervals. There was a time when we believed that biomes and their constituent species simply shifted up and down the continents, moving in concert from north to south and back again, neatly tracking the changing climate as the boundaries of climate zones shifted during the glacial epochs. However, an examination of fossil pollen records revealed that our original assumptions were wrong. Instead we found that individual species respond independently to changes in climate so that groups of species that occur together today did not necessarily co-occur during the geologic past (Davis 1987). For example, pollen records indicate that 12,000 years ago forests in present-day Ohio and Tennessee were dominated by spruce, oak, ash, hop-hornbeam and sedges, which formed a type of forest that does not exist anywhere in the world today, even though all of those species continue to survive (Fox 2007). Today the forests in Ohio and Tennessee are dominated by maples, oaks and hickories.

The past forests that have no contemporary equivalent are often referred to as no-analogue forests because no forests today are similar to them. Those forests came into existence due to a combination of environmental factors – such as temperature, precipitation and seasonal variation – that no longer existed, as well as the history of dispersal and establishment by individual species. It is increasingly obvious that no-analogue ecosystems exist all across the globe. They appear unusual to us because they have a species composition that is unlike any ecosystems we are familiar with. However, their unusual appearance is really just a consequence of our rather narrow frame of reference with respect to geologic time. We measure time in reference to human lifespans and many of us have a hard time imagining a time span of more than about five generations, centered on ourselves and ranging from our grandparents to our grandchildren. Maybe we have a slightly more expansive sense and can think across seven generations – great-grandparents to great-grandchildren, but beyond that our ability to grasp the immensity of time becomes a hindrance to us. In terms of geologic time, the ecosystems that exist today represent just a tiny, very recent snapshot of what has occurred at a particular location. One set of model results predicts that by 2100, anywhere from 4 percent to 39 percent of the earth's surface will have no-analogue climatic conditions. Existing climatic conditions will be lost from 4–48 percent of the earth's land surface (Fox 2007). It is highly likely that the newly developed climatic conditions will support ecosystems formed from combinations of species that have no analogue anywhere on earth today. So, if we can find past evidence of climate change and the formation and reformation of constantly changing ecosystems that are unlike any ecosystem we know today in terms of species composition, then why are we worried about projected climate change and no-analogue ecosystems in the future?

The answer and the reason we worry so much is because the rate and scale of climate change and formation of new ecosystems is so much greater than at any time in the geologic past (Millennium Ecosystem Assessment 2005; Mooney 2010). In the geologic past, species were limited in the distance and rate at which they could disperse by their dispersal mechanisms – for plants whether the species dispersed passively by wind or water vs. actively being carried by animals, for animals whether they dispersed by crawling, walking, swimming, flying or hitchhiking, etc. (Davis 1987). But today the rate and distance of dispersal for many species have been greatly accelerated and increased by human activities (Ricciardi 2007). The movement of species around the globe is absolutely unprecedented, resulting in the constant arrival of ecologically, previously unknown (for a given ecosystem) new species to practically every ecosystem on earth.

Novel, no-analogue ecosystems arise from a combination of factors – primarily land-use change and habitat destruction, climate change and the arrival of non-native, potentially invasive species (Hobbs *et al.* 2006, 2009). Those three factors interact to form a triple-whammy in which it is absolutely inevitable that currently existing ecosystems will change and that their new form will be different from anything we have seen before. Habitat destruction creates openings which allow the establishment of new individuals and species and tends to favor species with excellent dispersal abilities and rapid growth rates. Many non-native species that flourish are ones that tolerate human disturbance regimes and which disperse easily (often with human assistance) and grow quickly.

Rapid climate change means that local species may be at a disadvantage when disturbances occur, because the conditions which would most favor their establishment (and which were probably present when their parents became established there) no longer exist or at the least are no longer as favorable to them. The numbers of non-native species arriving in different parts of the globe are staggering. The flora of New Zealand has been especially well studied, and botanists have carefully documented all the native and non-native species that are established and growing there. New Zealand is home to 2,065 native plant species and 24,774 non-native plants, of which at least 2,200 have become naturalized and grow without human care (the remaining 22,000+ non-native species can only persist due to human cultivation) (Duncan and Williams 2002; Norton 2009). So now, non-native species account for over half the species of "wild" growing flora in New Zealand! There are more than 10,000 non-native species known to be established in Europe. At least 1,094 of those non-native species are known to have ecological impacts and at least 1,347 of them are known to have economic impacts (Vila *et al.* 2010). Non-native species may have positive effects on their new ecosystem, but more often, when they have an impact, the impact is negative.

The development of novel ecosystems throws a large spanner into our usual schema of attempting to restore and manage ecosystems within their historical range of variability (HRV) (Seastedt *et al.* 2008; Bradley *et al.* 2010). In the coming new norm for the twenty-first century, the concept of managing for the HRV is rapidly becoming an anachronism. There may be situations in which conditions remain similar enough to the recent past and present that we can talk meaningfully about restoring to incorporate the HRV. But it is likely that such situations will be increasingly rare. In fact, it is quite possible that the best case scenarios in the future will be the development of what some have termed "hybrid" ecosystems that have some of the characteristics of the current or historical ecosystem but which, due to changes in terms of species composition and function, now exist outside the HRV (Figure 5.1).

In the best case scenario hybrid ecosystems will still contain original keystone species and most of the original ecosystem functions, so restoration and management efforts will be best employed to maintain the hybrid ecosystem in good condition, supplying desired ecosystem services. Truly novel, no-analogue ecosystems are most likely to develop when conditions have changed so drastically that the original keystone species have been lost and many of the original ecosystem functions have also been lost or at

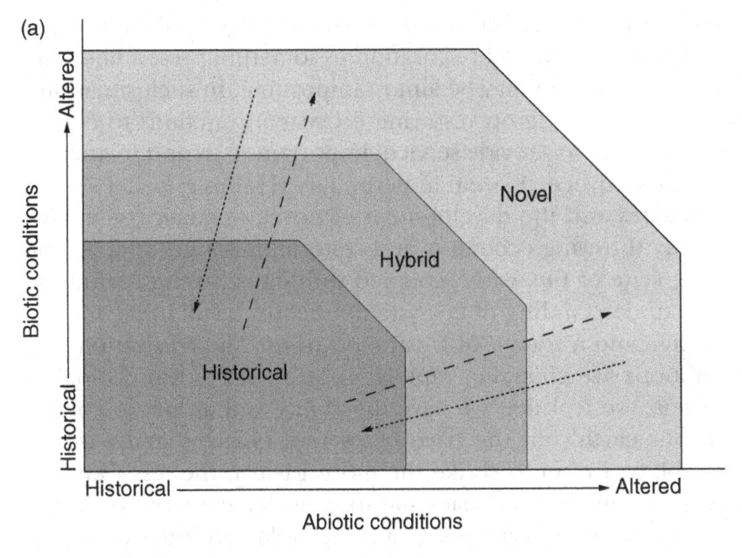

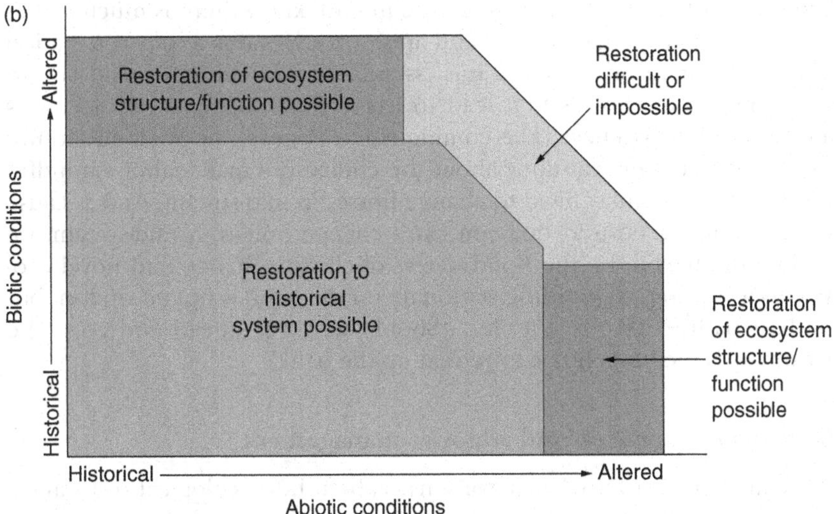

Figure 5.1 The top diagram (a) shows the current situation with respect to ecosystem change. As the physical environment changes due to climate change and other factors and the biotic environment changes due to invasive species, hybrid and novel ecosystems form. The dashed arrows pointing out (up) show the change from historical to hybrid and novel conditions. The dotted arrows pointing in (down) show what we attempt to do with restoration – move altered ecosystems back toward historical conditions. The bottom diagram (b) indicates that as physical and biotic conditions change it becomes more difficult or even impossible to return to historical conditions. Thus, in some instances, it makes more sense to restore ecosystem structure and function rather than historical conditions. If the situation is too greatly changed, it might not be possible to do any restoration (source: redrawn from Hobbs *et al.* 2009).

least highly altered. In those situations the ecosystem has probably crossed a threshold or tipping point, and restoration to something resembling the original, historical ecosystem may be almost impossible. In such situations, it may be best to concentrate on restoring ecosystem functions to ensure the ecosystem continues to provide services important to us and to attempt to maintain the new novel ecosystem in perpetuity (Hobbs *et al.* 2009).

Changing climates and the development of novel, no-analogue ecosystems are currently throwing ecologists and restorationists for a loop. The rapidly changing state of the world, coupled with our growing realization that our current understanding of ecosystem function is still in its infancy, is both a challenge and a source of frustration to us. The frustrations are obvious – conditions are changing rapidly, we are engaged in ecological restoration because we feel that we have the ability and a duty to correct past abuses of our earth, but the pace of current changes make us less certain of our ability. I often feel like the more I learn, the less I know – because every new thing I learn leads me to consider many more things that I don't know and have yet to learn, and the cycle continues endlessly. And that is where the challenge comes in – we know there is much still to learn, so we have to try to learn and apply those lessons to our restoration and management projects. But it is essential that we not get ahead of ourselves and make choices that lead to irreversible management decisions and restoration practices. The coming rise of a novel, no-analogue future should make us very cautious about the choices we make, always mindful of the fact that we may need to change horses in mid-stream. And because we have often been told that you can't change horses in mid-stream, we are in a double-bind – the flood waters of climate change and novel ecosystems are rising, so standing stationary on the bank is not an option, but the best path to take is not clear. Should we forge ahead and cross the stream, or is the best choice to retreat up the bank?

Ecological restoration and adaptive management

There are no clear answers or consensus about how ecological restoration will be used in the twenty-first century or about the role ecological restorationists should play as we respond to the challenges arising due to global environmental change. It is obvious that the world is changing, but we don't know how rapid or extensive those changes will be. Thus it is very difficult to decide on the best course of action with respect to ecological restoration in the next 100 years. What is obvious is that restorationists are experiencing considerable angst about the future. Many, if not most, restorationists became involved in ecological restoration because they felt it presented them with an opportunity to make a positive contribution to helping solve environmental problems. But at the current time the pace of change is so rapid and the state of the future is so uncertain that we don't know whether the approaches to ecological restoration that developed over the past 50 years will still be viable in the future. It appears to me, based on conversations with many restorationists and attendance at many conferences and workshops over the years, that the current uncertainty and the resulting angst are leading many restorationists to question what they are doing and why they are doing it. Because there is no consensus about the future of the field, ecological restoration is currently in a state of flux, and just below the surface there is considerable disagreement among restorationists – whether academics, professionals or lay practitioners – about the direction the field should take.

One option is to place continued emphasis on our traditional approach to restoration – i.e., that ecological restoration should be focused on restoring ecosystems so that they return a site to the ecosystem that occurred there in the pre-disturbance past. Ecosystems are not static, so the ecosystem must be managed to allow for dynamic changes, but the ecosystem should be maintained within the HRV so that it has continuity with the past and properly honors the original ecosystem we are restoring. Thus I might conclude that the tallgrass prairie restorations I manage in western Illinois should continue to be maintained as tallgrass prairies because they are the ecosystem that was there prior to disturbance and their restoration was the original goal of the restorationists working at Green Oaks.

Another approach is to look at the amount of global climate change that has already happened, look at future projected change and then ask what ecosystem we can reasonably expect to maintain at this site in the future. Because we cannot predict exactly what conditions will exist at a particular site in 100 years, we need to be flexible about what we restore and attempt to maintain, using restoration to add in a wide range of species that have the potential to survive and prosper at the site well into the future. The focus of ecological restoration then becomes not maintenance of an ecosystem based on pre-disturbance conditions and the HRV, but rather the focus shifts to using restoration to maintain as much biodiversity and as many ecosystems as possible, given expected climate change. If the ecosystem with the best chance of long-term survival is quite different to the one that existed there in the past, then we may have to accept that change is inevitable and do the best we can. So as I read predictions that tell me western Illinois is likely to have a far different climate than exists there today or existed there in 1820 at the time of Euro-American settlement, I have to ask whether maintaining tallgrass prairie at Green Oaks will be possible, whether the effort necessary to try to maintain that prairie is sustainable and effort well spent. If the answer is that it will be impossible to maintain tallgrass prairie, then I need to start doing something different. That is a radical departure from our traditional approach to restoration.

A third approach, not entirely separate from the second, is to use ecological restoration in a proactive attempt to mitigate the effects of global environmental change (Galatowitsch 2009). We have traditionally used ecological restoration to restore and maintain ecosystems as a way of repairing past damages. We often think of conservation biology as focused on preservation and protection of valuable ecosystems and biodiversity, while ecological restoration is focused on bringing back what was lost (Young 2000). In that tradition ecological restoration is a way of maintaining a neutral balance of biodiversity, ecological communities and ecosystem services. Using ecological restoration as mitigation in a proactive manner is essentially suggesting that we should work for a positive increase in ecosystem services. Such an approach will require land-use changes beyond our usual goal of restoring degraded or destroyed ecosystems and we must think of ways to put all ecosystems, even agricultural, urban and semi-natural, to work to help absorb carbon. Restoration as proactive mitigation of climate change is also a significant departure from our traditional way of doing ecological restoration.

In the twenty-first century, ecological restoration will be increasingly important as a tool to mitigate and repair the damage due to global climate change (Galatowitsch 2009). The United Nations Environment Programme developed a plan entitled Reducing Emissions from Deforestation and Forest Degradation (REDD), which is centered on ecological restoration of forests (UN-REDD Framework Document 2008). Forests have tremendous potential as living storehouses of CO_2 because of their ability to absorb huge amounts of CO_2 via photosynthesis. The REDD program alone will generate many opportunities for ecological restoration. Ecosystems restored via REDD and other climate change mitigation projects will have to be long-term restorations because to have a positive effect on reducing atmospheric CO_2 the forests have to grow undisturbed, without harvest, for at least 25 years. One hundred years of growth without harvest is even better (Galatowitsch 2009). It may be possible to restore forests on more than one billion hectares (an area larger than Canada). Studies by the Global Partnership on Forest Landscape Restoration have found that marginal, low-productivity agricultural lands in many countries have the best potential for restoration. If those lands are restored to forests the combination of improvements in agriculture and the fact that the land to be restored was marginal will result in no net loss of food production (Chestney 2009). Restoring forests on an area larger than Canada would provide tremendous opportunities to implement the REDD program goals and help mitigate climate change. Those restorations will almost certainly return many more ecosystem services than just carbon sequestration and will benefit large numbers of species, not just the trees being promoted as living carbon stores.

However, the REDD program is not without controversy. Indigenous peoples and local communities in developing nations have been skeptical about both local benefits of the REDD program's forest plantings and how the REDD program will affect local ownership and management of forests

(Danielsen *et al.* 2011). In particular there are fears that REDD will be administered by national governments that may not be inclined to pay much attention to local interests and concerns. To help address those issues, a new version of REDD known as REDD+ was approved at the 2009 Copenhagen meeting of the 15th Conference of Parties of the United Nations Framework Convention on Climate Change (UNFCCC). In particular, REDD+ represented an expansion of REDD so that although reductions in carbon emissions remained the primary goal of the program, the expanded REDD+ also included provisions for using it to promote both economic opportunities for local, indigenous peoples and the preservation of biodiversity (UNFCCC 2010). How would REDD+ promote economic opportunities and support the lifestyles of local people living in and near REDD+ projects? The UNFCCC document doesn't provide details, but one possibility that is frequently mentioned is hiring local, indigenous people to do the monitoring and management of the REDD+ projects.

Danielsen and colleagues (2011) performed a comparative study at forest sites in India, Tanzania and Madagascar in which the accuracy of local people and outsider-trained professionals at monitoring forests for carbon sequestration and biodiversity were compared. They found that local, indigenous people were as accurate as professionals and much more cost-effective because of lower labor costs and proximity to the sites being monitored. Therefore, supporting local communities via hiring indigenous people to do the monitoring and management is a viable option.

However, hiring local people to do monitoring and management does not end concerns about who will control the land and make larger-scale decisions about land use and long-term planning. Dennis Martinez, the founder and chair of the SER Indigenous People's Restoration Network worries that REDD+ will lead to continued loss of access to land and thus traditional lifestyles for indigenous peoples around the world (Martinez 2011).

Martinez's objections to REDD+ take two forms. First, he thinks it is far easier and less expensive for polluters to pay for forest preservation and aforestation than it is to actually reduce their carbon emissions; thus there is little incentive to reduce carbon emissions. Second, he feels that many REDD+ forest preservation schemes place high value on having trees growing in place, but little to no value in maintaining the traditional practices of indigenous people who live in those forests. He fears that many REDD+ programs will prevent indigenous people from using their forests and prevent them from maintaining their cultures and lifestyles. If REDD+ is to be truly beneficial to all people, the programs will have to maintain traditional land use by indigenous people. The fact that REDD+ may actually create a disincentive to reducing carbon emissions is a tougher nut to crack. If we rely on economic incentives to catalyze a reduction in carbon emissions, then the cost of releasing carbon will have to increase via taxes or fines to the point that achieving actual reductions in emissions becomes more cost-effective than continuing to release carbon while paying for carbon sequestration elsewhere. Thus far most governments have shown little interest in regulating carbon emissions via high levies for releasing significant amounts of carbon.

I am torn as I think both globally about the future direction of ecological restoration and locally as I worry about how to properly manage my tallgrass prairie restorations. In many ways I tend to be a traditionalist. I like bringing back the ecosystem that existed prior to disturbance. I like the idea of knowing the plant communities that my ancestors saw. There is something deeply satisfying about walking through the prairies, especially on an autumn afternoon when the asters, sunflowers and gentians are in bloom and the grasses are becoming golden-red as they go dormant, and thinking back about all those who walked through autumnal prairies before I did – Blackhawk and his people, Pere Marquette, Abraham Lincoln, Willa Cather – the list can go on and on. But then there are times, usually deep in the dog days of July and August with high humidity and searing temperatures, when I look up from another day of cutting down and poisoning the black locust and autumn olive that are trying so hard to prosper in my prairie and I ask myself why I bother. My struggles are Sisyphean, my victories seem Pyrrhic at best. Perhaps it is time to take another approach and ask whether I should start sowing the seeds of another ecosystem. But when I was hired, I was given clear instructions that I was expected to study and maintain the prairies at Green Oaks. So I shoulder my saw and spray can and I continue removing the brush and hoping for prairie resurgence. But I also think it is time to re-educate my college community about the long-term viability of the ecosystems at Green Oaks. We might all agree it is time to change course.

The question of how restorationists should respond to the projected scale of global change is vexing everyone in the field. Even the best minds among us do not have definitive answers. Eric Higgs suggests that we use "wild design" to plan restoration in the future, especially in natural areas such as national parks or nature reserves (Higgs and Hobbs 2010). For Higgs any sort of ecological restoration is an intervention and all interventions are the result of design. He claims that wild design explicitly acknowledges the interplay of human activity and ecological processes. When we intervene in the ecological world, our intervention imposes ethical responsibilities. Therefore wild design (and by extension restoration) must be based on the following core principles: clarity, historical fidelity, resilience, restraint, respect, responsibility and engagement. Design must be based on a clearly articulated approach to solve a particular problem. That solution should have fidelity to the ecosystem that historically existed at the site. However, the solution also has to be resilient to coming climate change and that, as we have seen, will be a huge challenge. Higgs realizes that ecological restoration in the future may well involve working with hybrid and novel ecosystems (he is a co-author on the Hobbs *et al.* 2009 paper discussing them), but he cautions that we must exercise some restraint and respect for the environment when planning restorations that use or promote novel ecosystems. He asks us to pursue a middle path between a slavish adherence to the traditional focus on HRV and moving completely in the direction of restoration for the sake of maintaining ecosystem services first and worrying about historical conditions as a distant second concern. How do we strike that balance and find the middle path?

For increasing numbers of restorationists the solution is to use ecological restoration as adaptive management (Folke *et al.* 2005; Hobbs and Cramer 2008; Hansen *et al.* 2010; Kiesecker *et al.* 2010; Lawler *et al.* 2010). Adaptive management is not a completely new approach. C.S. Holling first defined adaptive management and argued in 1978 that we should be using adaptive techniques. Adaptive management is usually understood by scientists to mean a process of experimental design and implementation of management that occurs simultaneously and continuously so that the process of learning about a system happens while the system is being managed. Project design, management, monitoring and re-evaluation are all brought together so that we learn as we are doing and thus can make changes to design and management in mid-course. The best pithy description I have seen is that "adaptation is a bicycle we must build as we ride it" (Hansen *et al.* 2010, p. 64), which sounds even more difficult than changing horses in mid-stream, but there you are – that is the world we live in today.

Thinking about ecological restoration as a form of adaptive management has some advantages because such an approach explicitly acknowledges that we understand the near future is going to be a time of rapid change and great uncertainty; thus any management plan we undertake has to be flexible. Folke and colleagues (2005) emphasized that adaptive management forces us to:

- learn to live with change and uncertainty;
- combine different types and sources of knowledge for learning how to manage effectively;
- create opportunities for self-organization so that both ecological and social systems are more resilient in the future;
- nurture resilience so that we are able to promote renewal, learning, growth and new forms of management.

One of the key strategies in adaptive management is making sure that human needs are addressed during ecological management. Adaptive management has been used most often by large policy-making groups charged with complex problems such as managing the Columbia River Basin in western North America so that the river system will continue to provide hydroelectric power and flood control while also preserving what remains of Columbia's salmon populations and restoring the salmon fishery in that watershed (McClain and Lee 1996; Folke *et al.* 2005).

Large environmental problems always have social dimensions and it is vital that human societies are included in the planning, management, monitoring and re-adjustment that occur during adaptive management. Folke and colleagues (2005, p. 445) understand adaptive management to be "about bringing together old knowledge, from diverse sources, into new perspectives for practice." Because human concerns are so central to adaptive management and because economic factors place limits on what we can achieve via management, various economic instruments such as cost–benefit analysis will be an important part of our design and monitoring. The main problem we have when making economic decisions about management issues is that our current economic system largely (actually almost completely) ignores the value of ecosystem services and thus they are vastly undervalued. Traditional cost–benefit analysis fails to capture the real value of ecosystem services, as well as the cultural importance of those services. Along with making use of adaptive management for our ecological restorations, global society has to undergo institutional change in order to more properly place values on ecosystem services and natural processes. We know that in human communities that place a high value on natural processes, such institutional change is possible. If communities do not place a high value on natural processes, then institutional change and adaptive management are difficult, if not impossible (Daily *et al.* 2009). Ideally, the process of adaptive management will help us gain more knowledge and a better understanding of the dynamics of the ecosystem we are managing, and greater ability to sustain and promote the resources and services from that system.

So can we conclude that adaptive management sounds like a great way to think of ecological restoration and move forward under the adaptive management banner? Perhaps we can, but not without considerably more discussion because adaptive management has proved to be a contentious issue among restorationists.

The first bone of contention involves questions of whether adaptive management really lives up to its promises of combining design, management, monitoring and re-evaluation to create workable solutions that satisfy human stakeholders. There is some evidence that early attempts at adaptive management were not entirely successful. This was at least partly due to adaptive management being applied to large, complex systems (like the Columbia River Basin or controlling spruce budworm across eastern Canada) that impacted large numbers of people, many of whom had different needs from the system and different ideas about how to best satisfy their needs. But the lack of success was also due to the fact that the models employed were often linear (as with many climate change models), so they failed to capture the rates and types of change. Scientists also have a tendency to not fully appreciate or even listen to non-scientific arguments about how to manage systems. Thus, during some applications of adaptive management, many stakeholders felt disenfranchised and the social end of the social–ecological dynamic was not satisfied, resulting in failure to adequately implement management plans and to change the plans when change became necessary (McClain and Lee 1996).

But another area of contention with respect to adaptive management is in many ways a negative response to the name "adaptive management" and the implications of that name. For some restorationists, adaptive management implies that we are simply folding up our tents and going home, just because the restoration game has become more difficult. If we say that we are adapting to climate change and now focusing on maintaining novel ecosystems in order to continue receiving ecosystem services, it feels like we are abandoning our traditional principles in two ways. First, we appear to be giving up the idea of maintaining historical fidelity and managing for the historical range of variability. Instead we are saying that if a particular new/novel collection of species starts growing successfully on our site, then we will accept that situation and go from there (which may be what we are doing in some restorations). Ecological restoration has always been difficult and has always required continuous work to remove unwanted species and promote desired species. The critics of adaptive management fear that we are rolling over to absorb drastic changes to historical ecosystems at the exact moment when biodiversity is most threatened, and at the time when we should be doing everything we possibly can to preserve biodiversity and restore pre-disturbance ecosystems.

The third area of contention is that adaptive management puts social concerns ahead of ecological ones. For many restorationists, the act of ecological restoration is primarily about returning a site to a "more natural" condition that existed prior to disturbance. The need to maintain historical fidelity and biotic integrity during restoration trumps social needs for that site (Gobster (2001) provides a nice summary of the conflicts involved with restoring a park on the Chicago lakefront, and he describes some restorationists as having little interest in social concerns). Under adaptive management we appear to be actively managing ecosystems for human benefit and not giving enough consideration to the needs of other species.

For me, adaptive management is a somewhat unfortunate term because of the history of marginal success with adaptive management and the implications that can be drawn from the name. However, I think the basic guiding principles of adaptive management – that design, management, monitoring and re-evaluation have to occur simultaneously and continuously – will be absolutely essential in all restorations conducted during the twenty-first century.

One of my favorite scenes from the Wallace and Gromit movies occurs during *The Wrong Trousers* when Gromit is in hot pursuit of the criminal penguin, Feathers McGraw. As they race around Wallace's house, Gromit hops on a speeding model train. Because Gromit gets shunted onto a branch line and the train is about to run out of track, he grabs a supply of rails and begins a program of furiously laying down track in front of the train, always just getting a new piece of track placed before the train and Gromit race over it. As he scrambles to lay the track he keeps one eye on Feathers McGraw and the other on approaching obstacles, always managing to lay the track at just the right place and moment to go from one room to another and to avoid colliding with pieces of furniture on his mad journey in pursuit of the penguin. Sometimes I feel like we restorationists are playing the role of Gromit, laying track as fast as we can, negotiating obstacles as they pop up in front of us, always in pursuit of a goal that remains just in front of us, never quite in reach. Although Feathers McGraw is apprehended in the end, it does not happen until Gromit and Feathers have a spectacular crash. Hopefully we can use the techniques of adaptive management to keep up with our target without any spectacular crashes along the way.

Managed relocation as a special type of restoration

The rapid rate of climate change presents unique challenges to restorationists. The rate of climate change is increasing and is already more rapid than originally predicted (Parmesan 2006; Hansen *et al.* 2010; Lawler *et al.* 2010; Mooney 2010). For many species, especially plants and others with poor dispersal abilities, the rate of climate change is likely to be more rapid than their natural rate of migration into new areas with suitable environmental conditions. This has led some restorationists to propose engaging in managed relocation (MR) (also known as assisted migration, assisted colonization and assisted translocation), in which humans actively move species from a location where they exist today to a location that is likely to have better environmental conditions for them in the future (Richardson *et al.* 2009; Minteer and Collins 2010). MR is a controversial idea, but I will first discuss the rationale and potential benefits before discussing the criticisms.

The primary goal of MR is to increase the probability that species will survive climate change by moving them from locations predicted to be unsuitable in the future to sites predicted to be suitable. Usually this will mean moving species outside their current distributional limits and creating hybrid/novel ecosystems as we add new species to existing ecosystems. Most of the focus of discussion of MR has been on species with poor dispersal abilities – especially plants – with the assumption that if we build the proper ecosystem, the more mobile species will come afterwards. It is clearly a much more aggressive approach to ecological restoration because rather than simply repairing damages to a particular site using species native to that site in the past, we are proposing to move species to completely new areas, sometimes to areas that are not damaged at all, in order to promote species dispersal and survival.

For some restorationists the pace of environmental change is so great that we cannot sit and wait for it to happen, because by the time the effects of climate change are obvious (i.e., conditions at the site are so changed that resident, native species are suffering declines in survival, growth and reproduction), it may be too late to save many species, some may even have gone extinct or have so few remaining members that it will be impossible to put together a large enough population of colonists to successfully establish the species elsewhere. Instead, we must be proactive, anticipate changes and work to beat those changes so that we have the best chance of preserving species, biodiversity and ecosystem services (Richardson *et al.* 2009). Even those in favor of MR recognize the need to be cautious in its application because it represents such a departure from traditional practice (Minteer and Collins 2010; Richardson *et al.* 2009). But even while being cautious, some still claim that MR should not be the policy of last resort, but rather one of several tools we use in ecological restoration in the next 100 years (Richardson *et al.* 2009). And, surprisingly (or not given many people's desire to do *something* about environmental change), MR is already being used in some situations such as establishing individuals of the rare conifer *Torreya taxifolia* in North Carolina, far from their current home in Florida, and the establishment of several new colonist populations of about a dozen species of trees in British Columbia in areas outside of their current distributional limits (Minteer and Collins 2010). A recent BBC news report described a 2011 project in which vendace (*Coregonus vandesius*) were moved by pack llamas from Derwentwater, Cumbria, UK to Sprinkling Tarn, a mountain lake 500 m higher in the Lake District, because of fears Derwentwater will become too warm to support the vendace. Because vendace are the rarest freshwater fish in the United Kingdom, saving them is important for preserving regional biodiversity (BBC 2011).

Somewhat similar to MR, Bradley and colleagues (Bradley and Wilcove 2009; Bradley *et al.* 2010) have discussed the need for "transformative restoration" in which non-native species are planted in new areas in order to replace native and non-native species lost due to the effects of climate change. They propose transformative restoration as a way to establish beneficial non-native species in order to prevent the establishment of harmful, invasive non-native species in disturbed ecosystems (Bradley *et al.* 2010). Their justification seems a bit circular to me because it can be difficult to determine whether non-native species are truly beneficial, or at least neutral, to a new environment, partly because some non-native species appear to be neutral for decades before becoming problems as conditions change (Elton 1958).

The problems that may arise from MR are numerous (Minteer and Collins 2010; Ricciardi and Simberloff 2009). Some of the problems are primarily ecological problems. At the simplest level the problems center on issues of whether the species will survive and grow in the new locations and the fact that establishing new populations of species in novel areas will be expensive and labor intensive. There are also questions about whether the current population, now being used to donate offspring to the colony, can afford to lose any additional individuals beyond typical losses to mortality and natural dispersal. It is possible that the newly introduced species will hybridize with species present at the new site and that genetic introgression will result in maladaptation for both colonist and the original native. If we are not careful, we may inadvertently introduce new diseases along with the new species. How well can we predict the effects the new species will have on their new ecosystems?

We have many questions about the wisdom of introducing new species to an ecosystem because there are many well-documented problems that arise when non-native species become established in a new ecosystem and far too many examples of non-native species becoming destructive pests in their new environment (Mack *et al.* 2000). There are also a series of ethical questions that result from discussions of MR. Are restorationists exhibiting dangerous levels of hubris as they play at being not just local ecosystem managers but planetary managers? We have created many global changes inadvertently and those are causing us tremendous problems – it is another matter entirely to take on the role of deliberately making global changes to ecosystems.

Does a move to MR represent a weakening in our resolve and a failure on our part to do enough to combat environmental change and preserve existing populations *in situ*? Perhaps MR is actually an excuse for our failures to do enough in terms of local restoration and preservation – we blew it at one location but engaging in MR allows us to feel good about trying to make amends elsewhere? For many years restorationists and conservationists have tried to convince the public of the need to preserve the integrity of existing local ecosystems and have warned of the dangers of introducing non-native species to new areas. Won't MR confuse the public about our approach? Will the public even accept the necessity and desirability of MR? Given this large suite of potential problems it is clear that if we engage in MR we can only do so after long discussion and careful consideration of all the ecological and social ramifications of this new approach (Minteer and Collins 2010; Ricciardi and Simberloff 2009).

When I was in graduate school, we used to say that there is a fine line that separates the cutting edge from the lunatic fringe. In the case of MR it is possible to see where that fine line lies. Some conservationists have been talking about a truly radical form of MR that they refer to as the "Pleistocene Rewilding" (PR) (Donlan *et al.* 2006). The basic premise of PR is that as modern humans spread around the globe, they caused the extermination of many very large vertebrates – the great extinction of Pleistocene megafauna. That point is fairly non-controversial – there is excellent evidence of extinction of many large vertebrates in North and South America, Australia, Eurasia and on large oceanic islands following the arrival of modern humans. Proponents of PR argue that ecosystems in places without their megafauna are now depauperate and lack many ecological interactions and ecosystem functions found in areas that still have abundant megafauna – primarily areas of Africa and some parts of Asia. They argue that we should establish new colonist populations of still-existing megafauna in those areas that lack them to re-establish lost ecological interactions. This reintroduction would generate new evolutionary potential, help preserve megafauna now in decline in their home countries and perhaps generate more funds for conservation and income for local people living near the reintroduction sites.

The proposed first site for large-scale PR is the western United States. Proponents suggest that we (re)establish populations of the Bolsan tortoise, wild horses and donkeys, cheetahs, lions, Asian elephants and camels in the Great Plains and desert southwest. They suggest that PR is feasible in other parts of the world too, such as Australia, Siberia, parts of Asia, Europe and large oceanic islands. And while a part of me likes the idea of cheetahs chasing pronghorn antelope across Wyoming or elephants wandering the boulevards of France (and thus giving life to the Babar stories), PR definitely feels like a proposal from beyond the fringe. It has all the potential problems of MR only with the added bonus of bringing in really big animals that typically have really big effects on local ecosystems. Given that western North America lost its megafauna about 13,000 years ago, the ecological interactions that existed there between megafauna and other species have long been lost. Resident species have evolved since the unfortunate departure of the megafauna and newly introduced non-native megafauna would represent a significant disturbance to existing ecosystems in western North America. PR would be an extremely expensive and labor-intensive undertaking. It would be much better to concentrate on spending the time, effort and money on preserving megafauna like cheetahs and Asian elephants in their home countries (Rubenstein *et al.* 2006; Oliviera-Santos and Fernandez 2010).

So, if PR is a trip to the lunatic fringe, is MR representative of the cutting edge, despite its potential problems? At this time MR raises more questions than it answers. But I suspect that in the near future, the rate of environmental change will become so great that, after additional testing and consideration, MR of species within their home continents will become a mainstream form of ecological restoration.

Unique opportunities and special challenges

As is abundantly obvious, the twin specters of global environmental change and the rise of novel ecosystems present us with many challenges, not just as restorationists, but as people in general. The rapid pace of change for both the global environment and the development of novel ecosystems make these challenges especially daunting, indeed frightening, in their scale and complexity and our uncertainty about how to respond to them. And yet we know that challenges and the approach of a potential crisis also create opportunities because in such situations we are forced to innovate if we are to survive and prosper (Folke *et al.* 2005). Failure to innovate, to respond creatively to new challenges often means that the crisis will wash over us like a great wave, threatening to sweep us out to sea or to leave us cold, wet and exposed if we somehow manage to grasp a rock from which we can begin anew.

The idea that challenges breed opportunities is such a truism that many political campaigns are built on words to that effect: "Yes we live in difficult times but we can rise to the challenge and emerge renewed, reinvented and better than before." President Barack Obama's inaugural speech of January 20, 2009 was filled with such rhetoric (as are many inaugural speeches). Unfortunately for Americans and President Obama the many challenges the United States and he faced remain. The challenges brought about by global environmental change are far greater than those faced by any one nation, and the community of restorationists and conservationists must seize this opportunity in order to effect positive changes to our relationship with the earth and to each other so that we can find solutions that work for all. As restorationists we must be wary of raising awareness of challenges that then remain unresolved, leaving us open to the opportunity of being barracked for setting unrealistic expectations about our ability to deal with the challenges.

At the most basic level, we need institutional change so that we begin to properly value the earth and all it provides to us. As has been well documented by many authors, our failure to recognize the true value of the earth and our environment (whether we are measuring value by ecological, ethical, social, cultural, economic or any other yardstick) is at the center of our current environmental problems. Our human institutions – whether governmental, political or financial – seem set up to find ways to disagree with each other, thus preventing the change necessary to re-evaluate our relationship with the earth and each other. Entrenched, uncompromising interest groups have prevented the United States from signing documents such as the Kyoto Protocol. Canada has recently decided to withdraw from the Kyoto Protocol due to similar pressures. Disagreements among nations resulted in very little being accomplished at the Copenhagen 2009 talks. We must find a way to move past these institutional obstacles. If nothing else, global environmental change may (and I have to say it is only a case of "may") force us to finally make significant changes in our behavior at individual, corporate and governmental levels.

The necessary institutional changes will have to begin with a fundamental shift in our values such that we begin to do "economics as if nature matters, and ecology as if people matter" (in the words of Aronson *et al.* 2007, p. 2). We need to develop a more proper, realistic valuation system for ecosystem services (Slootweg and van Beukering 2008; Daily *et al.* 2009;

Nelson *et al.* 2009). However, accomplishing such change has proved difficult and will continue to be difficult. Detailed models based on local ecosystem services have demonstrated that if we enhance biodiversity we also enhance the provision of ecosystem services. The same models have also demonstrated that at the current time there is a negative correlation between the values of commodities produced in a region and the preservation of biodiversity and ecosystem services (Nelson *et al.* 2009). If we rely on market forces alone, we are unlikely to be able to achieve large-scale preservation of biodiversity and ecosystem services. There will have to be the provision of subsidies or financial rewards to promote biodiversity and ecosystem services (Daily *et al.* 2009; Nelson *et al.* 2009). Peer approval is also a very strong incentive, so if a local ethic of "preservation of biodiversity and enhancing ecosystem services is a good thing" develops we are more likely to achieve the desired changes. Sometimes negative incentives such as legal sanctions or just making people feel guilty for not participating in the new management scheme may be necessary (Daily *et al.* 2009). But relying on negative enforcements should be a policy of last resort because it may generate hard feelings and be counter-productive. We know that policy-makers and stakeholders respond positively once they recognize the value of ecosystem services and restored ecosystems (Slootweg and van Beukering 2008). Restorationists, whether academics, researchers, professionals or lay practitioners, must be better at communicating the benefits and successes of maintaining and restoring ecosystem services (Groffman *et al.* 2010).

Scientists involved with restoration must become better at communicating and make more of an effort to reach non-scientists (Olson 2009; Groffman *et al.* 2010). It is especially important to recognize that people get their knowledge and form their worldviews within their social network. Most scientists live outside the social networks of the average person and are often surprised that not everyone thinks the way they do. This is at least partly because many scientists and professionals are somewhat peripatetic. We frequently grow up in one place, go to one or two different colleges and universities in different towns for our education and training, and wind up living and working in yet another place – and may well move from one institution to another during our careers. Our closest social relationships tend to be with other scientists and professionals. We have short-term ties to local communities and are sometimes looked at as uninformed outsiders who don't understand the local scene. I now live and work 15 miles from where I grew up in western Illinois. I am at least the fifth generation of my family to live here. But when I meet local people who don't know me, they sometimes think of me first as an outsider professor and are a bit suspect until they learn my family history. Where I live, it is important to have local connections in order to be fully (or at least easily) integrated into the social network. If you lack those local connections, you must work that much harder to become integrated. Therefore, to be effective communicators scientists need to make more of an effort to understand how those social networks work and how best to reach them.

The good news is that thanks to changes in technology, especially the development of online social networks and blogs, it is easier and easier to reach a large audience. Another piece of good news is that in general scientists are regarded as experts and trustworthy (especially compared to, say, politicians or salespeople) and people are eager to hear what scientists have to say (Olson 2009; Groffman *et al.* 2010). Thus, on-going environmental change provides us with a subject on which our expertise is desired, and technology provides us with better and better tools to reach a large audience. The bad news is that scientists are particularly bad at speaking in language that the average person finds interesting (Olson 2009). We need to learn how to speak to that audience and to reach members of the audience most likely to have influence on other members of the public.

Scientists often also think that their message has failed to reach the public because of a lack of education or information. They then attempt to fill the perceived gap in knowledge by supplying more and more information in more and more detail (often while being long-winded and boring) in such a way that they frequently overwhelm the public. The public is better informed than scientists sometimes think and the failure of the public to act in ways that scientists think is best is more often the result of how the message is delivered than the content of the message. The best approach is to pick the most important idea and present it in language that is interesting, quick and to the point (Olson 2009; Groffman *et al.* 2010).

At times scientists are condescending to the public, which is bad; but it is even worse to be insulting. Richard Dawkins may be a great scientist, but when he accuses religious people of being idiots, all he is doing is ending any chance of dialogue with them. His followers may gloat smugly and say "Right on Richard," but they will soon find they are left talking only to each other and not making any kind of progress. Ecological restoration involves direct interventions with nature and often requires changes in land use. For those who value wild nature it can seem like we are attempting to impose our will on the land. For land owners who now have to change their behavior, it can feel like outsiders are assuming they know better than the land owner how to treat the land. In either case, restoration can feel offensive and generate resistance (Swart and van der Windt 2010). I have encountered a general unease and resistance to restoration many times when talking to people about ecological restoration in both the United States and the United Kingdom. In order to use restoration effectively to counter climate change we must learn to engage the public in a positive way. Changing people's minds is hard and takes time. Turning them off is unfortunately much easier and only takes one bad encounter.

Restorationists face another set of challenges due to the legal and political environment in which they must work. At the local and larger scale, frequently the landscape is divided into smaller plots with many different land owners, and this pattern of mixed ownership complicates the implementation of many restoration plans (Ravenscroft *et al.* 2010). Different land owners may have different current uses for the land, different needs in terms of what kind of economic return they require from the land and different plans for its future use. In some instances the legal requirements for land use vary depending on the type of land owner, so that private individuals, corporations, government agencies and NGOs all have different regulations governing their use of similar landscapes. In this kind of ownership environment, it can be extremely difficult to establish coordinated restoration and conservation projects. We know that we achieve greater success in restoring both biodiversity and ecosystem services when we can restore larger pieces of land (Doyle 2008), so it will be vital that we find ways to coordinate efforts among many land owners and develop incentives to encourage cooperation among them as well.

At a larger scale, international treaties impose a conservation framework that both provides opportunities and sets limitations for ecological restoration (Cliquet *et al.* 2009). The member states of the European Union tend to be progressive in their environmental policies. Many people living within EU nations are interested in ecological restoration, especially as a tool that might help mitigate the effects of climate change. But restoration and conservation work within the European Union all takes place within a carefully developed and negotiated framework of laws and guidelines.

Natura 2000 provides the primary set of conservation guidelines in the European Union. The goal of Natura 2000 is maintaining and establishing connectivity between ecosystems and resilience of ecosystems and was the result of a forward-thinking planning process. However, the implementation of EU environmental policy and law has been focused primarily on the conservation of particular species and habitats of special interest. The main directive from Natura 2000 established a goal of no deterioration or loss of protected species and habitats. Protected species and habitats must be maintained in the condition they existed in at the time of protection, which is a rather static approach that may be difficult to maintain during the coming environmental change. For example, the Birds Directive mandates the conservation of certain bird species and the habitats they occupy – if species of interest are present at a particular site then nothing can be done that would cause their decline, and efforts must be made to maintain them and their habitat in the current condition. If species go extinct at a particular site – suppose a marsh in Belgium dries up due to increasing temperatures and declines in precipitation and thus the protected marsh birds die out – then Belgium could be subjected to large fines (millions of euros) for failing to maintain the birds and marsh, even though the forces causing habitat loss and extinction were beyond Belgium's control. Thus the current legal framework in the European Union, even though it was progressive when first written and adopted, is already somewhat outdated and may not be an adequate instrument to address the issues likely to arise during rapid environmental change. The policy and planning framework may have to be expanded to consider the EU as a whole, rather than working on a member state by member state basis, and to recognize the highly dynamic nature of ecosystems during rapid change (Cliquet *et al.* 2009).

During rapid climate change the challenges of planning and managing restoration and conservation projects that cross international boundaries will become increasingly complex and vitally important to solve. Ecosystems and species do not recognize international boundaries, and those boundaries often cut through ecosystems, creating complex management issues (Lopez-Hoffman *et al.* 2010).

North America would seem to present a fairly simple situation because the continent only has three nations – the United States, Mexico and Canada – with long, continuous borders between only two nations at a time. Thus it would appear to be fairly easy to set up transboundary agreements that would be favorable for conservation and restoration. Unfortunately that is not the case.

The border between the United States and Mexico is an especially difficult area in which to achieve compromises that will benefit ecosystems and species. In the United States there is a great amount of concern (sometimes manifesting itself as fear) about illegal immigrants entering the United States across the United States–Mexico border. Therefore the border is tightly controlled, including the erection of fencing that cuts through ecosystems and divides populations of wild species. For one case in particular, the small, vulnerable population of jaguars in southern Arizona, the erection of a border fence is likely to isolate the Arizona jaguars from Mexican jaguars, with negative consequences for the Arizona population (McCain and Childs 2008).

The United States–Canada border is less problematic, but even here the two governments often take different approaches to conservation such that predicted climate change is likely to engender different responses in the two nations, resulting in different management plans and ultimately different ecosystems developing on either side of the border. The potential for divergent outcomes has been modeled for the area around the Quetico Boundary Waters Canoe Area (Ravenscroft *et al.* 2010) and similar divergent management strategies and resulting ecosystems will likely occur all along the United States–Canada border. Some international boundaries present much greater difficulties than exist in North America – the Korean peninsula is one such case. Other international boundaries may occur between nations that work well together. But in any case, global environmental change is truly global and will require international cooperation. Nations will have to develop the means to identify mutual interests, shared services and ultimately mutual benefits from conservation and restoration (Millennium Ecosystem Assessment 2005). This has not been easy, and in the aftermath of Copenhagen 2009 I'm not terribly sanguine about our ability to identify mutual interests and benefits in the near future, but it is a goal we must continue to strive to meet.

The way forward: renewed restoration

Rapid environmental change and the rise of novel ecosystems are challenging the way we think about and understand ecological restoration. We restorationists find ourselves trapped between a rock and a hard place – strict adherence to restoration of historical ecosystems and maintaining them within their HRV is probably doomed to fail in the twenty-first century. And yet embracing the establishment of novel ecosystems and abandoning reference to the pre-disturbance ecosystem, leaving historical fidelity behind, does not feel right.

We face questions of how much intervention in ecosystem processes and development is reasonable? Do we know enough to promote the emergence of novel ecosystems as reservoirs of ecosystem services? Are the models of projected change good enough to allow us to make valid decisions about how current actions will be manifested in future ecosystems? Are we courting hubris and accusations of playing god if we adopt aggressive restoration programs based on incomplete knowledge? How do we communicate our current understanding of ecosystem change and plans to preserve biodiversity and ecosystem services to the general public? Even if we can communicate our plans in a coherent and compelling fashion, will the general public accept our conclusions and support our plans? If they accept our plans at first, how well will they accept the inevitable setbacks as we learn on the job? How will the public respond to the lengthy amount of time required before restorations begin to show the benefits we said would result from them? It is no wonder that the field of ecological restoration feels like it is in a state of flux and that there are undercurrents of disagreement and uncertainty among restorationists. We are wrestling with huge questions that lack obvious or easy answers. As much as we can, we will have to use ecological restoration in the twenty-first century to travel the middle path of respect for past ecosystems and the social and cultural meanings they hold, while also using restoration as a tool to promote the preservation and maintenance of ecosystem services in all of their manifestations – ecological, economic, cultural, esthetic, etc. (Harris *et al.* 2006).

We want (and need) to make a positive contribution toward mitigating climate change, and think we have the tools (especially forest restoration) to do so. The rapid pace of change often leaves us breathless, stressed and a bit afraid. Yet we as a group cannot afford to panic. I'm afraid that restorationists have essentially one chance to reach the public and governments and effect positive change in environmental planning and management. Despite the pressure (even if the pressure is self-induced), we have to make sure we get it right, because if we make a large mistake in our proposals (or maybe any mistake at all), we may never get another chance to contribute to the solution of the problems arising from environmental change.

In the twenty-first century we need to move to what I think of as "renewed restoration" – ecological restoration that is restoration along the middle path: restoration projects that recognize the importance of the past pre-disturbance ecosystems by maintaining as much historical fidelity as possible, while utilizing the reality of rapidly changing, no-analogue, hybrid and novel ecosystems in order to promote biodiversity, ecosystem services and human connection to the environment.

Case study: Green Mountain – restoration and improved ecosystem function

Do we have any models of how to combine the preservation of the historical ecosystem while adjusting for the rapidly changing new world? I don't think we have any that are exactly analogous to our current situation, but the restoration of Green Mountain on Ascension Island provides a road map that we may follow (Wilkinson 2004). Ascension Island is a young (approximately one million years old) volcanic island in the middle of the Atlantic Ocean. At the time of human discovery in the sixteenth century, the island had a species-poor flora of about 25–30 plant species. Early sailors introduced rabbits, sheep, donkeys and rats, which had negative effects on the native flora. By 1836, when Charles Darwin visited the island on his homeward journey aboard the *Beagle*, Ascension had become a largely barren, sparsely vegetated rock. Darwin, perhaps anxious to get home to green and verdant England, found Ascension to be ugly and harsh.

In 1843, Joseph Hooker visited Ascension and decided that what Ascension needed was a program of tree planting to establish vegetation on the island. Beginning in 1850, he and others began shipping large numbers of saplings to Ascension, most of which were planted high on the slopes of what is now called Green Mountain. Within 100 years a lush, species-rich and productive tropical forest had developed on the upper half of Green Mountain. As a result the soils on Green Mountain are increasing in depth, there is greater water retention and the significant growth of trees is capturing large amounts of carbon (Wilkinson 2004). The original native plants continue to exist in other locations on Ascension. Thus Green Mountain represents probably the best example of the positive possibilities of restoration leading to novel ecosystems and providing enhanced ecosystem function and services. We must hope that the majority of our restorations in the twenty-first century are equally successful.

Ideally, renewed restoration should produce ecosystems that Nigel Dudley (2011) has described as authentic ecosystems – ecosystems that self-regulate and contain expected levels of biodiversity and ecosystem function based on past history, current climate and geography. However, authentic ecosystems are not likely to be exact copies of past ecosystems and may include new, non-native species. If we can somehow thread our way between the rock and the hard place, then we will have arrived at true renewal and restoration that honors the entire system – human and ecological – for the benefit of all members. I will examine renewed restoration in detail in Chapter 7.

6 Geographical variation in attitudes to ecological restoration and why it matters

In this chapter I will:

- describe my perceptions of geographical differences in attitudes and approaches toward ecological restoration – in particular with respect to North America and Australia in contrast to Europe;
- provide background by discussing previous surveys examining the field of ecological restoration;
- analyze the results of a survey of recently published papers on ecological restoration;
- discuss the results of an online survey of restorationists;
- conclude by describing the similarities and differences in attitudes toward ecological restoration that exist in different regions of the world and what they tell us about future directions for restoration.

Introduction: why compare attitudes about ecological restoration?

I have been attending and presenting papers at conferences of the Society for Ecological Restoration (SER) for the past ten years. One of the things I have enjoyed most about those conferences has been the opportunity to meet ecological restorationists and restoration ecologists from all around the world and the chance to learn from them as I listened to their paper presentations. It has been even more enjoyable to go out to dinner and a drink or two after the day's presentations are over – it is in those informal, personal and often spirited (in a good, invigorating way) after-hours conversations that I have learned the most about how ecological restoration is practiced around the world and the reasons why people engage in ecological restoration. Early on in my attendance at the SER conferences I realized that my European colleagues spoke a different language with respect to ecological restoration than did my North American and Australian colleagues. And it was not just that the Europeans described what they were doing in different terms, it was also obvious that their initial assumptions were often strikingly different to the assumptions employed by North Americans and Australians.

The European restorationists usually said they were doing ecological restoration in order to promote and preserve biodiversity and also to ensure that ecosystems continued to perform valuable ecosystem functions. In recent years they have talked more and more about restoration as a way to promote ecosystem services. While my colleagues from North America and Australia would also talk about restoration as an important tool for promoting biodiversity, ecosystem function and services, they were more likely to describe the starting point for their restorations as a desire to return a damaged ecosystem to its original historical characteristics, or at the least to return the ecosystem to its historical trajectory, allowing for continuing dynamic ecological and evolutionary change within the ecosystem. Biodiversity, ecosystem function and services were often seen as very desirable by-products of a restoration based on the prior historical condition. The Europeans often spoke about the need to ensure that restored ecosystems had cultural value and that traditional cultural practices, such as certain grazing and hay-making practices that created particular landscapes, were maintained. At first the only North Americans and Australians I heard talk about cultural value or cultural landscapes were either native people or restorationists working with native people, although that is beginning to change and the importance of maintaining cultural landscapes appears to be becoming more mainstream in North American and Australian restorations.

I was puzzled by those differences in both description of and starting point for restoration. My own background and training was all based on the idea of returning damaged ecosystems to the pre-disturbance condition. Talking about restoration with almost no mention of a pre-disturbance reference condition seemed odd to say the least. I wondered whether the differences I heard as people talked about their restorations was just because I happened to be talking to people with extreme views on the subject – perhaps the differences I perceived were just an artifact of me talking to a small group of people who I found most congenial and fun to have a drink with at the end of a long day of listening to talks. But suppose the differences in approach that I sensed were in fact real differences, representative of the attitudes of most restorationists working in Europe, Australia and North America? Then how did those differences arise and what did they say about the current state and future of ecological restoration? The majority of ecological restorationists are from North America, Europe and Australia (e.g., 2010 SER membership data shows 81.6 percent of its members were from North America, 9.2 percent from Europe and 5 percent from Australia; perhaps a more representative sample comes from a large literature survey described later in this chapter in which I found that 57.2 percent of all authors of papers on ecological restoration came from North America, 25.1 percent from Europe, 8.6 percent from Australia, 3.5 percent from Latin America, 3.3 percent from Asia, 1.5 percent from Oceania and 0.8 percent from Africa). I met many people from North America, Europe and Australia and began to understand the approaches to restoration there, but I didn't have any feeling at all about how restorationists in Africa, Asia and Latin America approached ecological restoration. I became a student of the history and philosophy of ecological restoration as I tried to learn more about why people practiced restoration and why particular practices developed in different areas of the world.

I described my understanding and interpretation of the history of ecological restoration in Chapter 2, so I will not reiterate it here. The major points are that as Marcus Hall (2005) pointed out, from the beginning, even when using similar methods to address similar problems (e.g., restoration of mountain valley watersheds to prevent excessive erosion and flooding), Europeans and North Americans had different ideas about why restoration was necessary. Joan Ehrenfeld (2000) (as discussed in Chapter 2) claimed that restorationists with different ecological and taxonomic interests have dissimilar goals and employ divergent terminology to describe their work. Much of this difference is due to historical development of various branches of the ecological restoration community. Do these historical differences in approaches to ecological restoration persist today? If so, is there a geographic or professional component to the differences?

If I was going to answer my questions and develop a better understanding of attitudes toward ecological restoration and the reasons why people engage in restoration I needed to do some more systematic research than simply talking to friendly people at conferences and doing as much reading as possible. Therefore I conducted a two-part survey of the field. Part one was an in-depth survey of the recently published literature describing restoration projects. Part two was an online survey of restorationists in which I asked respondents to rate their reasons for doing restoration and also to provide written commentary describing their reasons in more detail.

It is important that restorationists understand the breadth of views and approaches within the field, if only to better understand what fellow restorationists are doing and why. It may be that a broad diversity of approaches to restoration is the sign of a field that has not achieved consensus about its goals and methods and that the diversity reflects a certain weakness in ecological restoration – a weakness in both theory and practice. On the other hand, a variety of approaches may instead reflect a robust field in which restorationists recognize that each situation calls for its own set of methods and that strategies applied to restoration in one place and time are not appropriate for another. I think the latter situation is a better descriptor and that ecological restoration is a robust, healthy field, but is my thinking an accurate assessment of the field? It is also obvious to me that the field is currently in a state of flux, fueled partly by our changing understanding of the rapidity and scale of global environmental change and partly by what is essentially a changing of the guard – many restorationists who began their careers in the late 1960s and 1970s are nearing retirement and a new group of restorationists with their own backgrounds and experiences are becoming increasingly active and moving into leadership positions within the field. Such generational shifts occur in all fields, but given that ecological restoration as an organized activity is quite young – the SER was founded in 1988 – now is the first time such changes have occurred at a large scale within the restoration community. Thus it seems vital that we develop an understanding of the current state of ecological restoration and determine whether we can identify new directions emerging from the field and its current, seemingly rapidly changing state.

Previous surveys and reviews of ecological restoration

Prior to doing a survey to examine attitudes toward ecological restoration, I checked to see whether other people had conducted similar surveys. While surveys of ecological restoration (and indeed of various scientific and policy fields in general) seems to be something of a growth industry, I could not find any recent surveys that examined ecological restoration by asking the kind of questions about restoration goals that I was interested in. Brudvig (2011) found that the number of papers published annually on ecological restoration has increased about 1,600 percent from 1995 to 2010. There also has been an increase in the number of surveys conducted of ecological restoration and related ecological management fields – enough so that Shanley and Lopez (2009) report managers and practitioners experiencing "survey fatigue." There were many surveys that focus on smaller subsets of the field – either in terms of type or location of restoration (e.g., Rowe (2010), who surveyed North America prairie restorationists to determine the most widely applied best practices for prairie restoration) – but only a few were designed to provide a broad overview of ecological restoration. After reading all the surveys I could find, I felt that my survey would be a unique examination of current attitudes toward and goals for ecological restoration, and thus a worthwhile addition to our knowledge. But first a quick summary of what previous broad-based surveys and reviews of ecological restoration, restorationists and management reveal about the field.

Setting goals, defining success

Ever since the beginnings of ecological restoration, a major issue has been the question of setting goals and determining the successful fulfillment of those goals. Given the disparate historical and geographical roots of ecological restoration it is not surprising that restorationists working in different times and places would have differing goals and measures of success.

The question that needs to be asked is whether any kind of consensus has arisen about goals and success.

Joan Ehrenfeld (2000) thinks that the varying fields of specialization for early restorationists lead to different branches of ecological restoration developing along similar but divergent pathways, each with its own unique emphasis. She identified four main traditions within the overall field. People working on the restoration or rehabilitation of severely degraded landscapes such as abandoned surface mines and spoil banks worked toward simply producing a functional ecosystem on a site that would otherwise be unproductive. I say "simply" as if such a goal would be simple to achieve, but in fact rehabilitating severely degraded ecosystems is exceedingly hard to do. Other early restorationists recognized some species and ecosystems such as North American prairies and temperate savannas were disappearing so rapidly that they might be lost forever. For them restoration was focused on returning altered landscapes to a previously existing condition as a way to save endangered species and ecosystems. In many countries wetland restoration is legally mandated because we recognized the value (often not until most wetlands were lost) of wetlands as important ecosystems vital for flood control, maintaining the water table and as habitat for many desirable species of fish and fowl. Thus wetland restorationists usually focused on restoring ecosystem functions and services. Finally, there were restorationists, mainly coming from a European tradition of land management, who recognized the necessity of thinking about restoration on a large scale and who realized that successful restoration would depend on restoring entire watersheds and landscapes. The restoration of landscapes requires thinking about ecological restoration at a higher level because any particular landscape or watershed is likely to contain several different kinds of ecosystems and may well encompass several political, governmental units and/or cultural groups who will all be affected by the restoration.

In order to determine whether restorationists set goals in a manner that corresponded to her historical analysis, Ehrenfeld (2000) surveyed 100 papers published in the journal *Restoration Ecology* from 1997 to 1999. She found that 25 percent of the restoration projects described in the papers focused on restoring species populations. Another 30 percent were focused on restoring ecosystems, landscapes and watersheds with no attention given to particular species. The remaining 45 percent of the projects had a variety of goals, often two or more goals for a particular project. After her analysis, Ehrenfeld claimed that restoration projects vary due to initial starting conditions and desired outcomes so that there is no single goal or even set of goals that can be applied to all restorations. She felt that restorationists need to be realistic about what is possible at a particular place when setting goals for a restoration project and that when working with the general public we need to be explicit about both the possible benefits and limitations of the proposed project.

Ehrenfeld thinks that the most successful restoration projects will focus on entire ecosystems, and in an earlier paper had suggested that restorationists should adopt ecosystems as the proper unit for restoration projects (Ehrenfeld and Toth 1997). Because every ecosystem, whether damaged by human activity or not, has its own history, each ecosystem restoration project has to be approached as a unique situation requiring its own solution. Thus, when we discuss reference conditions for the desired outcome for a restoration project, we should really be discussing a range of different possible reference conditions measured at many sites that encompass expected variation for that type of ecosystem. In this respect she is prescient, giving an early call for an ecosystem focus at a time when many fellow North American restorationists were more concerned with a return to historical conditions.

From its beginning, the leadership of the SER recognized the need to establish clear goals for restoration projects. Along with developing and revising (several times) a definition for ecological restoration, the SER also developed a set of guidelines by which we can determine whether or not a restoration has been successful. The most recent iteration of those guidelines was published in the *SER Primer on Ecological Restoration* (2004). The *Primer* also provided the most widely used definition of ecological restoration today (given in Chapter 1). Both the definition and the guidelines for determining success of a restoration were based on a distinction between ecological management, which was seen as concerned with maintaining an ecosystem in its current condition, and ecological restoration, which was intended to assist the recovery of an ecosystem so that it will no longer be damaged or degraded.

The *Primer* is an extremely helpful and influential document, but like many documents written by a committee it contains some interesting inconsistencies. The authors say the goal of ecological restoration is to return a damaged ecosystem to its "intended trajectory" (a curious phrase which seems to imply that ecosystems have intentions), back to the historical ecological and evolutionary pathway the ecosystem would have been on had it not been disturbed by human activity (Society for Ecological Restoration Science and Policy Working Group 2004, p. 2). However, the authors go on to acknowledge that the North American focus on restoration of historical conditions is not appropriate for places like Europe with highly species-diverse cultural landscapes or in developing nations where a return to historical conditions may be at odds with people's need to support themselves (Society for Ecological Restoration Science and Policy Working Group 2004, p. 2). They claim that an ecosystem is fully restored and the project has been completed when the restoration "contains sufficient biotic and abiotic resources to continue its development without further assistance or subsidy" (Society for Ecological Restoration Science and Policy Working Group 2004, p. 3). However, they also note that in some cases, due to problems such as invasive species, on-going climate change and/or continued human development in the surrounding area, the restored ecosystem may require on-going, long-term, essentially eternal human effort in order to maintain the ecosystem and prevent it from falling back into its degraded state (Society for Ecological Restoration Science and Policy Working Group 2004, p. 2).

In a way, I feel like the authors were hedging their bets by allowing for a wide range of possibilities in terms of restoration goals and variation in the need for continuing maintenance. And I know from talking with Eric Higgs that the authors intentionally tried to include every possible restoration scenario in order to accommodate the broadest possible range of restoration practice within the definition and qualifications. Ultimately I like the broad range of definitions because it fits the reality of restoration as practiced around the world. Based on what I have seen, every situation is different and it would be impossible to develop a rigid definition that would apply to more than a very small subset of restoration projects. The tallgrass prairie and savanna restorations I work with are all based on a return to historical, pre-Euro-American disturbance, but all require perpetual management and human intervention to maintain them on the desired ecological trajectory. Without regularly applied fire, those ecosystems would soon become dominated by many woody species and grow into a woodland lacking prairie or savanna characteristics. Thus even individual restoration projects frequently have inconsistencies between stated goals and actual restoration practice.

In order to allow us to more easily visualize what a successful restoration would look like, the *SER Primer* listed nine attributes of a restored ecosystem. The authors noted that many restorations will not display all nine attributes and that some are exceedingly difficult to measure, but they felt that nonetheless, a successful restoration will have most of these nine attributes evolving in the desired direction when the restoration work is completed. According to the *SER Primer*, a restored ecosystem:

1. will have the species found in the reference ecosystem(s);
2. should be made up of mostly native species;
3. will have all functional groups of species present;
4. will have a physical environment that will support the species and entire ecosystem;
5. will have normal ecosystem function (relative to reference ecosystems);
6. will be integrated into the surrounding landscape;
7. will have had biotic and abiotic threats reduced or completely eliminated;
8. will be able to withstand and recover from the expected range of ecological stress; and
9. will be as self-sustaining as the reference ecosystem.

The obvious question – have any ecological restoration projects actually achieved all nine attributes? – was asked by Maria Ruiz-Jaen and Mitchell Aide (2005). In order to answer the question, they surveyed every paper published from 1993 to 2003 in volumes 1–11 of the journal *Restoration Ecology*. In their survey they only used projects that involved either spreading seeds or planting plants to initiate restoration. This is an interesting choice on their part because it left out projects that make physical improvements to the environment and then rely on natural recruitment of native species to populate the site – a method that has been successful in several parts of the world. Nonetheless, they found 468 projects that fit their criteria. They found most of the work was done in North America (53 percent of the projects), with many projects in Australia and Europe (19 percent and 16 percent, respectively) and relatively few in Africa, South America and Asia (4 percent, 4 percent and 3 percent, respectively). Sixty-one percent of the projects were focused on one taxonomic group, usually plants. The taxonomic groups of most interest in these projects were plants (79 percent) and arthropods (35 percent). Almost 60 percent of the projects were carried out in response to a legal requirement to restore damaged ecosystems. However, of the 468 projects they examined, only 68 explicitly attempted to measure the success of the restoration once the project was completed. None of the projects evaluated success in terms of all nine attributes mentioned in the *SER Primer*. Most of the projects examined success in terms of three main ecosystem attributes – species diversity, vegetation structure and ecological processes. Curiously, none of those ecosystem attributes corresponds exactly with the nine attributes listed by the SER. Species diversity is the main concern of both attribute 1 and 2 in the *SER Primer* (the restoration should have species found in the reference ecosystem and these should be mostly native species). Vegetation structure is a component of the third SER attribute (all functional groups present). Ecological processes are contained within the fifth SER attribute (functions normally). Given that the projects analyzed were all conducted before the *SER Primer* was published, it is perhaps not surprising that none examined all nine attributes, but it is a bit surprising that none of the measures of success identified by Ruiz-Jaen and Aide correspond exactly to the suggested SER attributes. There appears to be some disparity between the practice of ecological restoration and the SER vision of an ideal restoration project.

It is also worth noting that Ruiz-Jaen and Aide (2005) found a large degree of taxonomic bias in the focus of restoration projects, with most effort directed at restoring plants and arthropods. The emphasis on plants is not surprising because plants are the primary producers in almost all ecosystems and often have somewhat poor dispersal abilities. Restorationists frequently seek to establish plants first, with the assumption that if the plants are present, then other taxonomic groups will follow and become established on their own. Plants are effective ecosystem engineers whose presence creates and changes ecological function for the entire ecosystem. This is sometimes referred to as the "field of dreams" method of restoration – if we plant it, the others will come. And sometimes the field of dreams method works.

Arthropods are perhaps a less obvious focus, but arthropods function on many trophic levels, have very high diversity and are fairly straightforward to sample, so they are a reasonable group to focus on. Taxonomic bias is a fact of life in biology. In general, biologists have tended to study vertebrates much more frequently than would be expected based on the relatively small proportion of vertebrate species compared to other taxonomic groups. This bias is not terribly surprising given our interest in species similar to us – both because of similar anatomy and physiology and because we identify with species we think of as closely related to us. Conservation studies and conservation projects also have a taxonomic bias, but interestingly it is quite different to the bias observed in restoration projects. For example, vertebrates are far over-represented in conservation research relative to the number of vertebrate species (69 percent of published papers versus 3 percent of named species). Also, birds and mammals predominate among vertebrates, despite being less abundant than fish, reptiles and amphibians. Plants make up about 20 percent of conservation research, considerably less than their prevalence in restoration projects. To a large extent, conservation research follows conservation funding, which is mostly focused on birds and mammals because the public, which supplies the funds for conservation, is more interested in birds and mammals than other taxonomic groups (Clark and May 2002). But the difference between taxonomic emphasis in restoration and conservation may also represent the different origins of the two fields. Many early restorationists had backgrounds in botany, forestry or range management, whereas many early conservation biologists came from zoology and wildlife management.

Truman Young and colleagues (2005) reviewed the published literature on ecological restoration in order to answer the question of how much impact ecological theory had made on the practice of ecological restoration. They found that many concepts developed as part of the science of ecology were familiar to restorationists. They listed several ecological concepts that show up frequently in papers on restoration or in conversation with restorationists. Those concepts were: competition, niches, succession, recruitment limitation, facilitation, mutualisms, herbivory/predation, disturbance, island biogeography, ecosystem function, ecotypes and genetic diversity. They noted that questions about how biodiversity and functional groups influenced ecosystem function were of special interest to restorationists and were especially essential concepts to be understood when planning successful restoration projects. Young and colleagues also observed that restorationists are interested in historical contingency, with how yearly variation in conditions can have a profound effect on the outcome of a restoration. The fact that restorationists are aware of and interested in concepts of biodiversity and the role of functional groups should not be surprising given that many restorationists see establishing biodiversity and ecosystem function as vital goals. However, it is not clear that restorationists regularly make reference to concepts from ecological science when developing plans for a restoration project.

Moore and colleagues (2009) conducted a Delphi survey of eight experts in the field of ecology to ask them about how much impact current ideas on ecological theory had on applied areas of ecology such as environmental management, conservation and ecological restoration. A survey of only eight experts seems a bit limited, but the authors started by sampling the ecological literature to find experts who were widely and frequently published and cited and identified a group of 20 experts to survey but only eight actually participated in the entire Delphi survey process. So even though experts were identified, approached individually and informed they were respected experts whose opinion was desired, a majority of those approached did not participate – which fits in with Shanley and Lopez's (2009) observation about survey fatigue.

The experts who did participate thought that compared to the past, there is now greater understanding of ecological principles among managers and practitioners and an increasing focus on ecosystem processes when designing management, conservation and restoration programs. But the experts also noted that many practitioners were primarily interested in species and species diversity as the most practical way to approach their work because preservation of biodiversity is a good strategy when promoting management, conservation and restoration. A focus on biodiversity is also seen as practical to implement when designing a management program. The experts thought that practitioners should begin working at the landscape level in order to fully encompass the large-scale and ecosystem connections that are necessary to maintain both biodiversity and ecosystem function. Again, the experts did not explicitly frame their discussion in terms of goals for ecological restoration, but many areas they recommended emphasizing – preservation of biodiversity, ecosystem function and processes, and working at a landscape scale – fit in with broad goals adopted by many restorationists.

Focus on biodiversity

Many restoration projects are explicitly designed to enhance and preserve biodiversity. In fact, some restorationists claim that ecological restoration is the best way to enhance biodiversity in ecosystems that have been modified or damaged by humans (Brudvig 2011). The question then arises as to whether these restorations have actually resulted in significant increases in biodiversity.

Jose M. Rey Benayas and colleagues (2009) conducted a clever meta-analysis of published papers providing assessments of ecological restoration projects. They employed a restrictive requirement when determining whether to include projects in their analysis. Each assessment had to provide data that allowed a comparison of biodiversity at the restored site, an undamaged reference site and a degraded but unrestored site. It is unusual to find studies that incorporate measures at all three types of sites – it is particularly unusual for restoration ecologists to include data on degraded but unrestored sites as they usually just make a comparison between restored and reference sites. Rey Benayas and colleagues were able to find 89 papers that met their stringent requirements and provided data that would allow them to compare among restored, reference and degraded sites. Luckily, these studies encompassed a wide range of geographical locations and types of ecosystems.

Rey Benayas and colleagues found that restorations are successful at enhancing biodiversity, but not as successful as we hope. Degraded ecosystems had 51 percent of the biodiversity found in the reference ecosystems and restored ecosystems had 144 percent of the biodiversity in the degraded systems. But the restored ecosystems only had 86 percent of the biodiversity of the reference ecosystems. Therefore, the restored ecosystems were clearly better than the degraded ecosystems in terms of biodiversity, but did not contain as much biodiversity as the reference systems. Even when examining long-term restorations after a decade or more of restoration, the restored ecosystems did not contain as much biodiversity as the reference systems. Restorationists have been more successful at enhancing biodiversity in terrestrial than in aquatic ecosystems and also more successful at enhancing biodiversity in tropical than in temperate ecosystems. The latter result surprised me because for many years ecologists have feared that complex, highly species-rich tropical ecosystems would be especially fragile and unlikely to recover from human disturbance. But here is an indication that tropical ecosystems can be restored somewhat successfully. Other recent studies have found that secondary forests develop fairly quickly in at least some tropical systems and contain a large amount of biodiversity and preserve much ecosystem function (Chazdon 2008; DeClerck *et al.* 2010). Tropical ecosystems may be much more robust and resilient than we originally thought. While this is good news as we attempt to restore disturbed tropical lands, that news does not provide us with license to carelessly damage existing tropical ecosystems under the assumption that we can easily repair the damage. As we see from the Rey Benayas study, even fairly old and well-established restorations lack the biodiversity contained in undisturbed reference systems.

Sometimes sites with minimal restoration actually achieve higher levels of biodiversity than do sites that have received large amounts of resources and labor during restorations (Prach *et al.* 2001a, 2001b; Tischew and Kirmer 2007). Can we find some relationship between biodiversity and restoration methods that helps explain these somewhat unexpected results? To answer those questions, Lars Brudvig (2011) conducted a wide-ranging survey to find out what characteristics of ecological restoration projects were correlated with improvements in biodiversity. He performed an ISI Web of Science search of every paper published from 2000 to 2010 in the journals *Restoration Ecology*, *Ecological Applications* and the *Journal of Applied Ecology*. He found 1,314 papers that had the words "restoration," "rehabilitation" or "recreation" in them and from that sample randomly selected 300 papers. Of those 300, 276 covered ecological restoration and from those he selected 173 papers that described active restoration projects and 17 that provided models of restoration projects. From his final sample of 190 papers, he found that 97 percent of the restoration projects investigated how various site-specific factors influenced the biodiversity of the restoration and 78 percent found there was a correlation between site-specific factors and biodiversity. Only 11 percent of the papers examined how landscape-scale factors influenced biodiversity and a mere 4 percent of the papers investigated the continuing influence of past land use on the successful restoration of biodiversity. Sixty-four percent of the papers described terrestrial restoration projects and 63 percent were focused on plants. From this Brudvig concluded that at least some of the variability we observed in the success of restoring biodiversity is most likely due to the fact that many restoration projects are focused on local sites and do not consider the wider landscape in which the restorations exist. Frequently, restorationists engage in the previously mentioned "field of dreams – if we build it they will come" approach in which we set up the proper physical conditions and then count on natural dispersal supplying the site with native species. Even more frequently, we use a modified, more labor-intensive version of the field of dreams in which we hope that if we plant it they will come, so proper physical conditions are established and seeds are sown, plants are set out and we hope all the other species will soon arrive on their own. But if the regional species pool has been degraded at the landscape scale, or if there are barriers to dispersal even when the proper species are present, such approaches will result in a limited amount of recovery of biodiversity. In either case, Brudvig claimed restorationists need to think bigger, at the landscape scale, when planning and performing their restorations. If restorations are properly situated in a larger landscape with either undegraded sites or other restorations nearby so that an interconnected web of ecosystems is created, then we are much more likely to be successful when restoring biodiversity.

Focus on ecosystems and their services

Throughout the development of the field, there have been restorationists who are primarily interested in restoring entire ecosystems because of the benefits to be derived at the ecosystem level (Ehrenfeld 2000). And there have always been some restorationists who think that the ecosystem is the most appropriate level to focus on when planning for and assessing ecological restoration (Ehrenfeld and Toth 1997). Again, we can ask how successful ecological restoration has been at the ecosystem level. Rey Benayas and colleagues (2009) also assessed our success at restoring ecosystem services in their meta-analysis of 89 restoration projects. Their results examining ecosystem services were similar to their results examining biodiversity – restored ecosystems did provide significantly improved ecosystem services, but not as much as we had hoped for. They found that degraded ecosystems provided only 59 percent of the ecosystem services provided by undegraded reference ecosystems. Restored ecosystems returned 125 percent of the ecosystem services provided by degraded ecosystems, but restored ecosystems only provided 80 percent of the ecosystem services that reference ecosystems provided. As with their results for biodiversity, restorationists had more success increasing ecosystem services in terrestrial ecosystems than in aquatic ecosystems and in tropical ecosystems than in temperate ecosystems.

Focus on plant interactions

Restorationists often select plants as the taxonomic group of most interest because successfully establishing plants is usually vital to the success of any restoration project. When restorationists think about ecological interactions among plants during restoration, they are often most interested in competitive interactions because negative competitive interactions, especially with undesirable or invasive plants, are a potential threat to successful restoration (Allison 2011). However, there is increasing evidence that positive, facilitative interactions (in which one species enhances the growth of another) among plants are also important during restoration. There is even evidence that in some restorations positive facilitative interactions are more important than negative competitive interactions (Gomez-Aparicio 2009).

Lorena Gomez-Aparicio conducted a meta-analysis of the restoration literature in order to determine how well understood the role of plant species interactions are during restoration and how widespread is the study of such interactions. She did her meta-analysis by searching the ISI Web of Science for papers that combined mentions of plant species interactions with a restoration study. In order to be included in her study, the paper had to provide quantitative data that could be used in further numerical analysis, discuss research done in the field and provide measurements of target and neighbor plants. Overall she found 62 cases that examined the effect of facilitation on emergence, 287 cases that examined survival and facilitation, 202 cases that looked at growth and facilitation and 123 cases that examined density effects and facilitation. She found that most studies were of short duration – usually only for two years or less. Few studies lasted longer than five years. The short duration is a common feature and weakness of many restoration and conservation studies. The short duration of many ecological studies is often related to the length of research grants (usually three years or less) and the fact that much research is done by graduate students who have to complete their research in a fairly short period of time. She found that facilitation was most important and provided the largest benefit for the emergence of seedlings of target species and their survival. Other plants often had neutral or negative effects on the target species with respect to later growth and density. Facilitation was most beneficial in stressful arid/semi-arid and/or low-nutrient environments, and facilitation was less beneficial in more humid, less stressful environments such as wetlands and moist temperate ecosystems. On the one hand that result is not surprising, but it does clearly identify areas in which restorationists should be looking for ways to enhance the benefits of facilitative plant interactions. While Gomez-Aparicio was satisfied with the breadth of case studies she was able to find, she did call for increased study of the role of facilitation in ecological restoration and also the need for more long-term studies. Restorationists have been interested in species interactions from the very beginning, but looking for the presence of facilitation is a fairly new area for restorationists. Facilitation is not explicitly mentioned as a major attribute of a successful restoration in the *SER Primer*, but it would fall within the goal of restoring normal function, if facilitation is a typical feature of the restored ecosystem.

Focus on practice and outreach

Ecological restoration in its modern form has developed from a combination of academic researchers, environmental managers, restoration professionals, land owners and lay practitioners. This combination has created a robust, dynamic field characterized by many approaches and viewpoints. Communication among these groups is essential because so much actual restoration work is done by lay practitioners and volunteers, but most of the conceptual and theoretical advances originate with the academics.

It has been obvious to me for many years, almost as long as I have been working in ecological restoration, that the various groups don't always communicate well and sometimes they don't seem to understand each other. Academic researchers sometimes despair because they feel practitioners don't design their restoration projects well enough to allow for the collection of meaningful data, which makes it more difficult to evaluate restoration success. They also fear that practitioners rely on traditional or trial-and-error methods and don't incorporate the latest or best science when conducting their research. On the other hand, practitioners sometimes think that academics spend too much time and effort conducting experiments that test methods the practitioners already know work (or sometimes know don't work) and feel academics should spend more time getting on with actual restoration rather than mucking about with experiments. Many managers and professionals are somewhere in the middle, hoping for more useful advice from academics and more rigor from practitioners. I have often wondered whether others have noticed the same issues of poor communication or whether my own perceptions are simply an idiosyncratic reading of conversations I have had with different groups of people.

Patricia Shanley and Citlalli Lopez (2009) surveyed a group of researchers working in environmental management and conservation in order to find out how the researchers share the knowledge they gather and who they share it with. Their survey was not a random sample. The survey participants were identified by looking at members of various research partnership groups and by identifying collaborators with those groups. Participants were then interviewed by a variety of methods – in person and phone interviews, an e-mail survey and through list serves. They were able to survey 268 people from 29 countries. Their survey was unusual in that the largest number of participants came from Africa, followed in order by participants from Latin America, Asia and North America. Unfortunately they did not list how many came from each location, but most surveys tend to be top-heavy with participants from North America and Europe, so they have found an interesting and unique sample group. Eighty percent of the participants were either academics or researchers and the rest were managers, practitioners or independents. Most had many years of experience – 65 percent had worked in conservation for more than eight years and 21 percent had worked in conservation for more than 21 years. The participants were most interested in communicating the results of their research with other scientists and researchers. They felt that was the most important audience for their work and that publication in scientific journals was the most important measure of their performance as researchers. They reported that the institutional and peer-review process places an emphasis on quantity of publications and rapidity of publication – so it was important to do studies that were short-term and could be published quickly. Most of the researchers admitted that they thought publication in scientific journals was a poor way to reach policy-makers and the general public, and that such publications were also unlikely to be successful at promoting conservation or the development of conservation programs.

However, they felt very little incentive to reach out to non-scientists because they were not rewarded for such outreach and thus outreach did not seem very important for their careers. Clearly such a situation is not conducive to good communication between academics and others involved in restoration. (Although Shanley and Lopez were examining researchers involved in conservation and environmental management, I will make the not too outrageous leap of claiming that academics involved with restoration ecology are under similar pressures to publish in scientific journals and have a similar lack of incentive to engage in outreach to the public). As Shanley and Lopez note, "Exclusion of public users of natural resources from scientific research results is not an oversight, it is a systematic problem that has costly ramifications for conservation and development" (Shanley and Lopez 2009, p. 535). As a systematic problem, it cannot be corrected by simply hoping the problem will go away or even by the efforts of one or two people. There must be structural changes in how performance is measured and the need for outreach. The SER has made many excellent recommendations about the need for outreach and maintains a valuable website (The Global Restoration Network); also, the SER's flagship journal, *Restoration Ecology*, requires that each manuscript provides a summary of the study's "implications for practice." But a society like the SER does not evaluate people for hiring, tenure or promotion; it can only make recommendations about what its members should do. Institutions that actually hire people have to believe that outreach is valuable for it to be promoted.

Sadly, such a lack of outreach is not unique to scientists. Other academic fields suffer from the same restricted viewpoint that the only people worth talking to are people in the same field. Ecological restoration and conservation occur in a socio-political environment, and it would seem that political scientists would be logical allies as restoration ecologists and conservation biologists attempt to reach out to government policy-makers. However, political scientists pay very little attention to biodiversity and conservation (Agrawal and Ostrom 2006). As of 2006, addresses by the previous 12 presidents of the American Political Science Association, in which they outlined areas where new political science research was needed, all failed to mention conservation or biodiversity. Political scientists are often most interested in studying problems at the level of the nation-state, but conservation and restoration problems are often located at smaller regional or subregional scales. Just as in conservation science, the incentives within the field of political science favor political scientists working in areas of traditional concern to political science, so there is little incentive to examine questions important to other disciplines (Agrawal and Ostrom 2006).

One of the main goals of the SER has been to promote interactions between restoration ecologists and practitioners of ecological restoration, especially to encourage the exchange of ideas between the two. Robert Cabin and colleagues (2010) surveyed the participants at the 2009 SER World Conference in Perth, Australia in an effort to determine whether a healthy relationship and productive exchange of ideas between researchers and practitioners actually happens. They invited 536 people to take the survey, of whom 381 completed at least part of the survey. The majority of the respondents were from Australia or the Pacific (62 percent). A further 16 percent were from North America and 10 percent from Europe. Half of the respondents were involved with research, the others consisting of practitioners, managers, professionals and policy-makers. Respondents were asked to evaluate the state of both the science of restoration ecology and the practice of ecological restoration. Most (80 percent) felt the science was in good shape, that we have learned a lot and are making progress on understanding the mechanisms of restoration, but many (45 percent) felt the practice of ecological restoration was not currently adequate.

From the comments made by respondents it was obvious there was a rift between at least some scientists and practitioners, with some scientists thinking that practitioners are "undisciplined and overly reliant on uninformed gut feelings" (Cabin *et al.* 2010, p. 784). And at least some practitioners think that scientists do not have a good understanding of or experience with ecological restoration as performed in the "messy real world" (Cabin *et al.* 2010, p. 784). Only 26 percent of the respondents thought that restoration ecology and ecological restoration were mutually beneficial, while 53 percent felt restoration ecology and ecological restoration were occasionally mutually beneficial and occasionally independent and irrelevant. The rest felt they were not mutually beneficial. The biggest limitation to both good science and practice was a lack of funds. Scientists tended to think that the gap between scientists and practitioners was due to a lack of scientific knowledge and that poor education limited the development of ecological restoration. The scientists also tended to think the science was limited by the still-young and developing state of scientific knowledge.

Unfortunately my perceptions of a divide between academics and practitioners appear to be valid. I first came across that divide back in 1994, when I was planning an experiment to examine mechanisms of seedling establishment in tallgrass prairie restoration. I was excited about the possibilities of the research and was talking about it with the owner of a fairly large prairie seed production company. She just rolled her eyes and made a comment along the lines of "We already know all that" and then muttered something under her breath to a colleague about scientists always trying to reinvent the wheel.

At the time I was new to the field and thought to myself, "Well that is an unfortunate attitude. We have so much to learn." But now I have seen many restorations that were based on little more than gut instinct and trial and error, and in many cases those restorations have flourished – the plants have grown, there is high diversity where once there wasn't, arthropods and birds have arrived. The sites are flourishing. And I wonder, what value can scientists add to the process (a lot) and, more important, how can we add it? We must do a much better job of communicating with our colleagues in practice by better explaining what we have learned through rigorous scientific study. I remember hearing a seminar when I was in graduate school in which the speaker (unfortunately I can't remember who it was) said "Statistics introduced rigor to science; unfortunately it also introduced mortis." Sadly our insistence on reporting the exact levels of rigor sometimes turns off a non-scientific audience (Olson 2009). We can also greatly improve communication by carefully listening to and valuing what practitioners have learned from spending many, many hours in the field.

Literature survey of geographic variation in goals for restoration

Now that I have provided some background about past surveys on ecological restoration and conservation, it is time to return to my own survey. As I mentioned at the beginning of the chapter, I was curious about the differences I perceived between approaches to ecological restoration as practiced in North America and Australia versus such practices in Europe. One way to approach that question is to examine the published literature to see what people say in print about their reasons for conducting ecological restoration.

I am most interested in the current state of ecological restoration, so I focused on recently published papers. I used the SciVerse Scopus database to search all issues of the journals *Restoration Ecology*, *Journal of Applied Ecology*, *Ecological Applications* and *Environmental Management* published from the beginning of 2006 to the end of 2010. I chose these journals partly because they all publish a fair number of papers on ecological restoration, partly because they are based in different geographic regions: *Restoration Ecology* is based in two regions – North America due to its sponsorship by the SER (offices in Washington, DC) and Australia because its editorial offices are in Perth); *Ecological Applications* in North America thanks to its sponsoring society, the Ecological Society of America; and the *Journal of Applied Ecology* and *Environmental Management* based in Europe due to their publishers, the British Ecological Society and Springer, respectively. I also chose these journals partly because they each have slightly different target audiences and thus reach a fairly wide spectrum of researchers working in ecological restoration. I used the search terms "restor* OR rehabilit* OR recreat*" to capture articles discussing restoration, rehabilitation or re-creation of ecosystems. I found a total of 1,028 articles using those terms (503 in *Restoration Ecology*, 169 in *Journal of Applied Ecology*, 150 in *Ecological Applications* and 206 in *Environmental Management*). In order to be included in my survey, the article had to have a description of a restoration project that provided an explanation of the goals for the project and enough detail to allow me to score it for several characteristics – location of the project, location of the authors, the type of ecosystem restored, the taxonomic focus of the restoration, the kind of analysis conducted by the authors, whether the project included stakeholder involvement, whether an analysis of economic benefits of the restoration was included, as well as an explanation of restoration goal(s). I excluded editorials and book reviews from consideration. In the end I found 677 papers that fit my criteria (427 from *Restoration Ecology*, 100 from *Journal of Applied Ecology*, 67 from *Ecological Applications* and 83 from *Environmental Management*). Using the term "recreat*" in my search terms yielded a fair number of extraneous studies that were focused on recreation as in recreational activities like fishing, camping, etc. and not on the re-creation of previously damaged ecosystems, but even so I was able to get a good crop of articles to review.

I performed statistical comparisons of the data collected by using contingency tables and the log-likelihood ratio (also known as the G test). I used the log-likelihood ratio both because the computations are simpler and because in cases when there are cells in the contingency table where the observed minus the expected value is less than the expected value, the log-likelihood ratio is considered as more robust than the Chi-square test (Zar 1999). The calculated value of G is then compared to values in a standard Chi-square table in order to determine whether to accept or reject the null hypothesis. Thus when I report the results of the tests I will give value of G calculated and the relevant Chi-square critical value.

For the most part, authors work with restorations located on the continent on which they live (Figure 6.1); however, there is a significant difference between author location and location of the restoration (G=39.56, critical $\chi^2_{0.05(7)}=14.067$, $p<0.001$). This difference is due to a significant

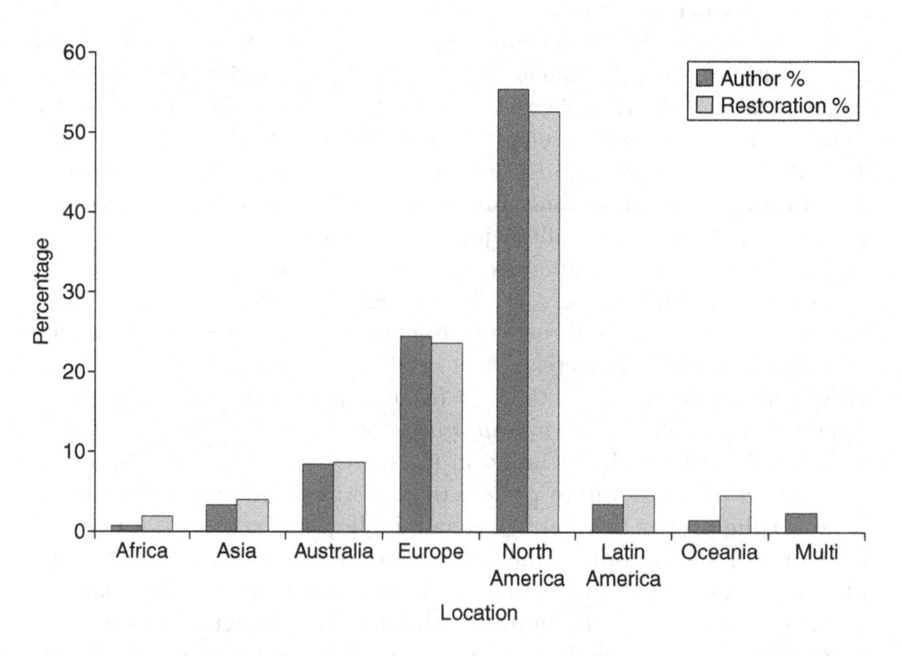

Figure 6.1 Percentage of papers written by authors from major geographical areas versus percentage of restoration projects located in major geographical areas.

Notes

Multi refers to papers written by many authors in which the authors came from different continents. There was a significant difference between location of authors and location of restoration projects (G=39.56, critical $\chi^2_{0.05(7)}=14.067$, $p<0.001$). Please note that the log-likelihood ratio was performed on count data but the graphs show percentages to allow more equitable visual comparison among geographical areas.

number of authors from North America and Europe working with restoration projects in other parts of the world – Africa, Asia, Latin America and Oceania. There were also a few papers with authors from several continents – such as Europe, Australia and North America – who reported on restoration on just one continent – perhaps Australia. By far the largest number of restoration projects were located in North America (357 out of a total of 677), followed by Europe (160), Australia (58), Latin America and Oceania (31 each), Asia (28) and finally Africa (13).

As I read the papers I scored them in terms of the type of study reported in the paper (Figure 6.2). I described the most frequent type of study as being "both" because those papers reported the results of both the progress of a restoration project and the result of an experiment conducted at the restoration site. A few papers simply reported on the progress of a restoration and were labeled as restoration papers. Some papers presented the results of an experiment conducted as part of a larger restoration but did not provide an analysis of the overall progress of

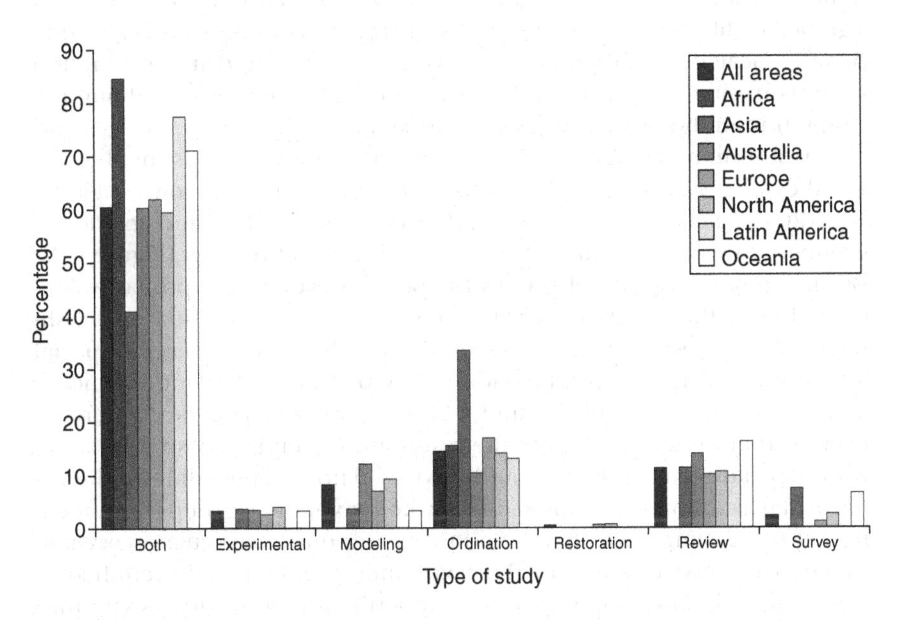

Figure 6.2 A comparison of the types of study of the restoration projects reported by geographic region.

Notes
Study types are described in the chapter. There were no significant differences among types of study by geographic region (G=49.92, critical $\chi^2_{0.05(36)}=50.998$). Data for all areas show the combined total for all the regions and are included to show how individual regions differed from the overall group. Please note that the log-likelihood ratio was performed on count data but the graphs show percentages to allow more equitable visual comparison among geographic areas. Please also note that the data for all areas were not included in the statistical tests so as not to test the same data twice.

the restoration. I called those papers experimental and included them in my survey if they provided a description of why the restoration was done. There were several modeling papers which presented models designed to predict the likely outcome of restoration. These models often were tied to specific restoration projects with explicit goals, so I included them in the survey. The papers that I described as ordination papers included both true ordination studies and studies using other correlational analyses, but in all cases provided data on a large number of variables being compared between restored and reference sites (at the least – sometimes they included data from several such sites or added in damaged, unrestored sites). There were a few survey studies in which people either living near or directly involved with a restoration were surveyed to determine their response to the restoration. And finally I found some papers that reviewed the progress of several restoration projects and provided an analysis of the reasons for the restoration. I found no significant difference in the types of studies published about restorations in different regions (G=49.92, critical $\chi^2_{0.05(36)}=50.998$). There is a tendency for more ordination, correlational studies to be published about Asian restorations, but it is not a significant difference. This result was intriguing because it indicates that people located in different regions are at least studying restoration projects in similar ways, even if they have different reasons for doing the restoration in the first place (as we shall see).

If restoration ecologists are doing similar kinds of studies in all areas, are they also studying similar kinds of organisms in their own regions? Broadly speaking the answer is yes, they do focus on the same taxonomic groups with relatively similar levels of interest in all regions (Figure 6.3). By far the largest group of studies (45 percent overall) are primarily concerned with the plants involved in restoration – their initial establishment, later growth and patterns of species diversity all receive special attention. The next largest taxonomic group of interest is not technically a taxonomic group at all. I found a large number of papers (25 percent overall) that were interested in the restoration of entire ecosystems along with their attendant functions and services. Those papers did not discuss particular taxonomic groups at all, or when taxonomic groups were mentioned it was simply for their role in carrying out a particular ecosystem function or service. Some of the taxonomic groups that I identified at first seem to be overlapping, but they describe what the authors said they were interested in. Thus some studies were interested in the restoration of all animals or vertebrates or invertebrates as a large class of organisms, but other studies were much more specific and focused on particular groups of animals, especially birds, insects and mammals. There were no significant differences in the taxonomic group of most interest to restorationists among the different geographic areas (G=99.0, critical $\chi^2_{0.05(84)}=106.395$). There is a slight tendency for plants to be of even more interest in Africa and Latin America and for those two regions to have less work done at the level of entire ecosystems, but it is not a significant difference.

In many ways the focus on plants is not surprising and fits in with what other surveys of the state of ecological restoration have found. Ruiz-Jaen and Aide (2005) and Brudvig (2011) both reported that most restoration literature was primarily concerned with plant restoration. As noted before, this kind of taxonomic bias is not unusual or unique to ecological restoration (Clark and May 2002; Young *et al.* 2005). The overall interest in plants certainly reflects both the origins of the field among restorationists who were interested in plants due to initial training in specialties like botany, plant ecology, forestry and range management, and also the history of establishing plants first because they are (relatively) easy to get established and we often assume the other species will follow.

I should not be surprised at this bias in the current literature, but I did think that restorationists were moving beyond the field of dreams strategy

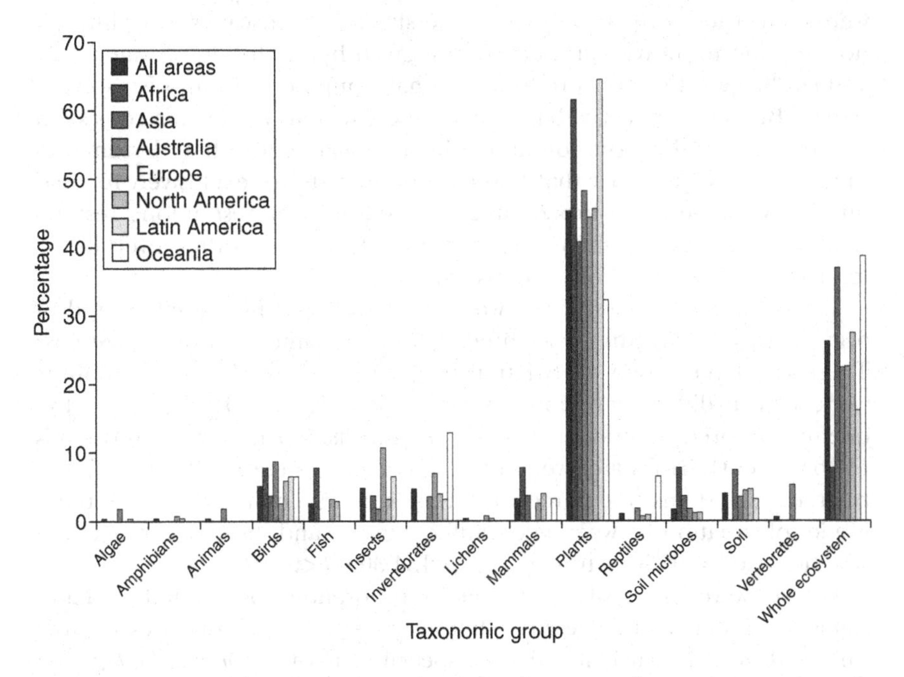

Figure 6.3 A comparison by region of taxonomic group of most interest in each restoration project.

Notes

There were no significant differences in taxonomic focus among the different regions (G=99.0, critical $\chi^2_{0.05(84)}=106.395$). Data for all areas show the combined total for all the regions and are included to show how individual regions differed from the overall group. Please note that the log-likelihood ratio was performed on count data but the graphs show percentages to allow more equitable visual comparison among geographic areas. Please also note that the data for all areas were not included in the statistical tests so as not to test the same data twice.

of "if we plant it they will come" and beginning to think more holistically about restoration. And there is some evidence of this change because both Ruiz-Jaen and Aide (2005) and Brudvig (2011) reported an even greater bias toward plants, with 61 percent and 63 percent of their survey studies, respectively, being primarily concerned with plants, compared to my 45 percent. Ruiz-Jaen and Aide were looking at papers published from 1993 until 2003, while Brudvig looked at papers published from 2000 to 2010, so both had an older dataset than I examined (papers published 2006–2010).

Further evidence that restorationists are moving from an almost obsessive focus on plants (and I say that as someone who is mostly interested in plants) is that about 25 percent of the studies were concerned with the restoration of entire ecosystems. While some branches of ecological restoration have always been interested in ecosystems, it is hard to quantify how large that branch was in the past. Ehrenfeld (2000, p. 3) reported that 30 percent of 100 papers published in *Restoration Ecology* from 1997–1999 were concerned with "ecosystems, watersheds or landscapes, with little or no attention to individual species." But given her inclusion of watersheds and landscapes, I'm not sure how well that compares with my own scoring system. But it is safe to say that a focus on ecosystems has been a consistent part of ecological restoration and at least appears not to have lessened in the recent past. The fact that fewer restorationists are exclusively focused on plants may reflect a broadening of taxonomic interest among restorationists so that there is greater recognition of species at other levels in the food web and the interconnections among them.

If restorationists around the world are interested in the same kind of organisms, then are they also interested in the same kind of ecosystems? There is a large range of variation in the ecosystems that restorationists work with in different regions of the world (Figure 6.4). The most frequently restored ecosystems overall are forests (30 percent), grasslands (17.6 percent), freshwater wetlands (9.2 percent), streams (9 percent) and riparian ecosystems (8.6 percent). But, as can be seen in Figure 6.4, there is a lot of variation between geographic regions, and those differences are significant ($G=303.5$, critical $\chi^2_{0.05(108)}=134.369$, $p<0.001$).

Forests were especially prevalent in restorations performed in Latin America, Australia and Oceania. The Australian emphasis on forests at first surprised me, but partly it reflects a special issue of *Restoration Ecology* that reported on several restorations of Jarrah forest in southwestern Australia. Africa was especially rich in restorations of savanna and shrublands. Agricultural ecosystems were most commonly restored as functioning agricultural areas in Europe. Coral reefs and streams were of special interest in Oceania. Brudvig (2011) reported that 64 percent of his papers were concerned with terrestrial ecosystems. I found about 25 percent of the studies I reviewed were restorations of aquatic systems; the rest were all terrestrial ecosystems. Ehrenfeld (2000) found 15 percent of the projects she reviewed

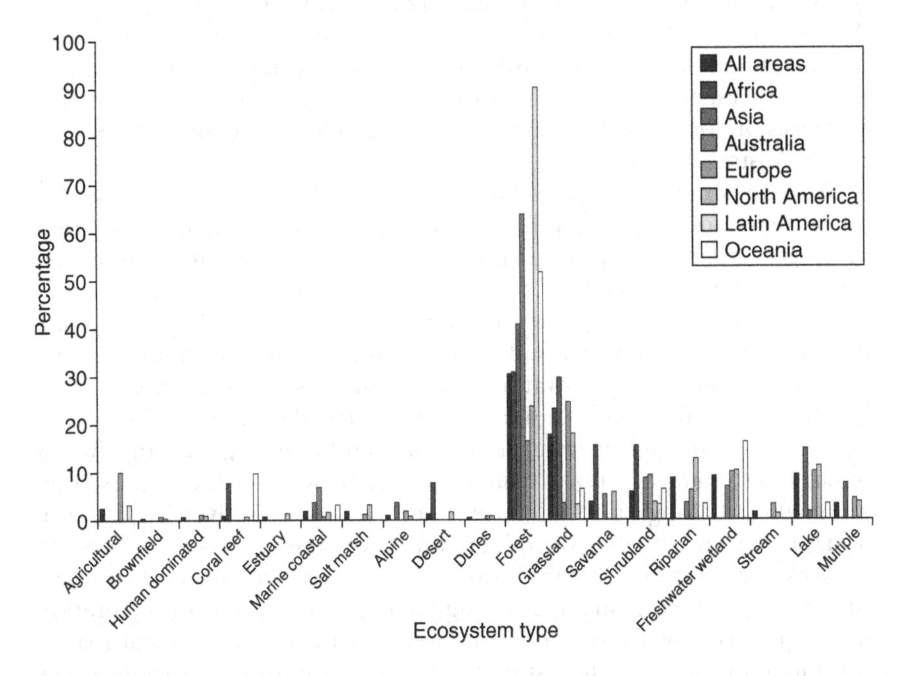

Figure 6.4 A comparison by region of the ecosystem type that was restored in each restoration project.

Notes

There was a significant difference among regions with respect to the type of ecosystem restored (G=303.5, critical $\chi^2_{0.05(108)} = 134.369$, $p<0.001$). Data for all areas show the combined total for all the regions and are included to show how individual regions differed from the overall group. Please note that the log-likelihood ratio was performed on count data but the graphs show percentages to allow more equitable visual comparison among geographic areas. Please also note that the data for all areas were not included in the statistical tests so as not to test the same data twice.

were located on mined or other types of severely damaged ecosystems, 18 percent were wetland restoration projects and the rest were about evenly divided between forests, grasslands, aquatic and other types of ecosystems – so about 19 percent for each of her categories. I did not score ecosystems exactly the same way that Ehrenfeld did – if mined or severely degraded sites were restored to an ecosystem similar to what existed there prior to disturbance, I scored it as restoration of that type of ecosystem (e.g., if there was a forest that was strip-mined and then a forest was planted I called that a forest restoration). If, on the other hand, a severely degraded site was restored to a completely different ecosystem maintained for human benefit I scored that as a human-dominated ecosystem (e.g., former grassland was destroyed and then replaced with a wetland designed to capture urban run-off). Ehrenfeld found evidence of more wetland restoration and less forest restoration than I did. Restorationists working in different regions usually expend most effort restoring the ecosystems available to them – e.g., coral reefs are restored more often in Oceania – but sometimes restorationists work hard to restore ecosystems that are either unusual or especially endangered in their region – such as the Australian emphasis on forests, while there were no examples in my survey of restoration of Australian deserts.

My main interest was to learn whether restorationists working in different parts of the world have different goals for their restoration projects. As I read through the papers I found that the authors frequently listed more than one goal for their restorations, which makes sense because often we are trying to accomplish several, hopefully compatible tasks with any particular environmental management scheme. Therefore I recorded restoration goals as the first or primary goal, that which seemed to be most important for a particular project, followed by a second goal of slightly lesser importance and a third goal with still less importance – but still enough importance to be worth achieving. I scored the goals as belonging to one of nine categories – biodiversity preservation/enhancement, climate mitigation, ecosystem function, ecosystem services, restoration of the historical ecosystem, control of invasive species, ecosystem/landscape management, development of restoration techniques (really a tool for restoration, but for many a critical reason for doing a project was to test methods) and a grab bag of various other reasons that included such things as cultural preservation, creating commercial landscapes and fisheries enhancement. My first look at restoration goals was done by combining all the goals – first, second and third – together in one overall dataset (Figure 6.5). In the overall dataset, examining all areas together revealed that biodiversity enhancement was the most frequent goal (23.9 percent), followed by restoration techniques (19.4 percent), historical ecosystems (16.2 percent), ecosystem functions (14.9 percent), management (13.8 percent) and ecosystem services (8.6 percent).

I was initially surprised by the number of papers that reported testing techniques as a goal for the restoration, but the more I thought about that emphasis, the more it began to make sense. Many of the restorations were young, small scale and were being written about by restoration ecologists primarily interested in developing better restoration techniques – so naturally they would design restorations in such a way that they could test those techniques. There were significant differences among regions with respect to restoration goals ($G=187.68$, critical $\chi^2_{0.05(48)}=65.171$, $p<0.001$). Australian and North American restorationists were much more interested in re-establishing historical ecosystems (the ecosystem present prior to disturbance) than were people in other parts of the world – Asian restorationists appeared to be particularly uninterested in historical restorations. Asian restorationists stand out for being especially interested in ecosystem services and ecosystem function. Asian, European, Latin American and

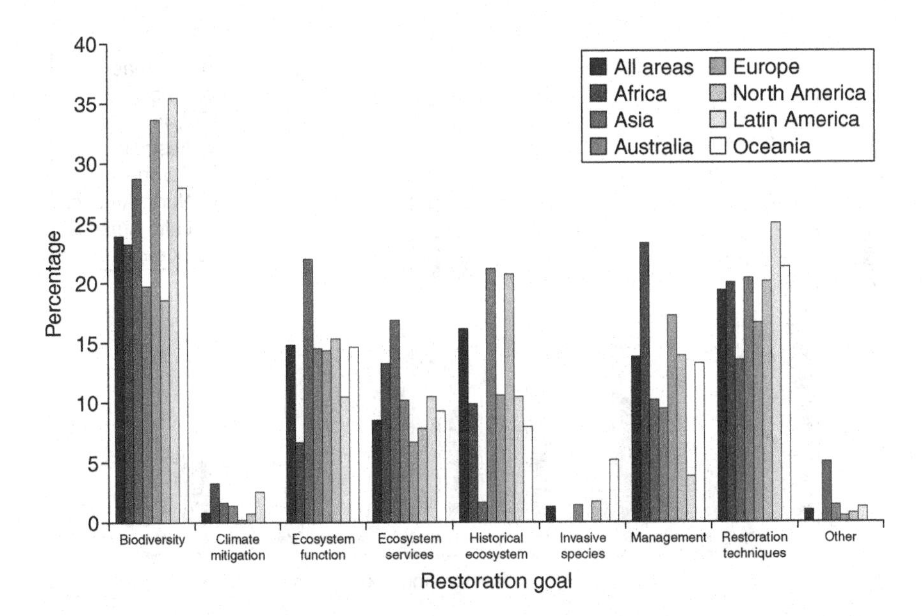

Figure 6.5 Up to three restoration goals were recorded for each ecological restoration.

Notes
In this figure the goals are all combined for an overall picture of restoration goals identified by restorationists in each geographical region. The combined goals for ecological restoration projects were significantly different among the regions (G=187.68, critical $\chi^2_{0.05(48)}=65.171$, $p<0.001$). Data for all areas show the combined total for all the regions and are included to show how individual regions differed from the overall group. Please note that the log-likelihood ratio was performed on count data but the graphs show percentages to allow more equitable visual comparison among geographical areas. Please also note that the data for all areas were not included in the statistical tests so as not to test the same data twice.

Oceanian restorationists all emphasized enhancing biodiversity. Management was an especially large concern for African restorationists.

When I examined the first or primary goal, the patterns were similar. Historical ecosystems (32.7 percent) and biodiversity (29.2 percent) are the most frequent primary goals for restoration (Figure 6.6). This is partly due to the significant regional differences (G=129.42, critical $\chi^2_{0.05(48)}=65.171$, $p<0.001$) where Australian and North American restorationists were much more focused on historical ecosystems than were restorationists from other regions – Asian restorationists not choosing historical ecosystems as a primary goal for any of their restorations. Biodiversity ranked especially highly for Asian, European, Latin American and Oceanian restorationists. Ecosystem function and ecosystem services ranked highly for African and Asian restorationists. Management and restoration techniques were not particularly important as a primary goal for most restorationists.

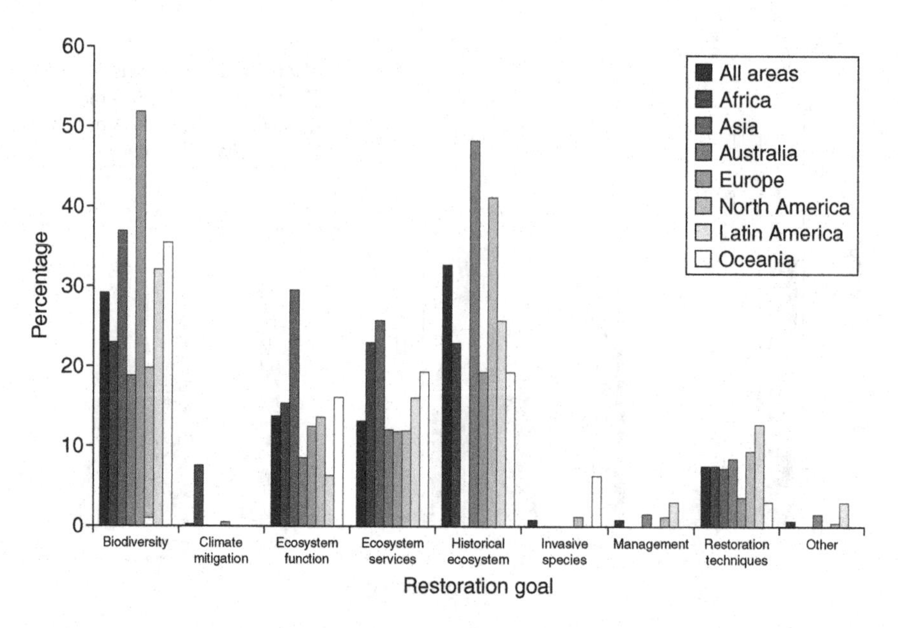

Figure 6.6 A comparison by region of the primary goal for each restoration project.

Notes

There was a significant difference in the primary restoration goal among the geographic regions (G=129.42, critical $\chi^2_{0.05(48)}=65.171$, $p<0.001$). Data for all areas show the combined total for all the regions and are included to show how individual regions differed from the overall group. Please note that the log-likelihood ratio was performed on count data but the graphs show percentages to allow more equitable visual comparison among geographical areas. Please also note that the data for all areas were not included in the statistical tests so as not to test the same data twice.

So what does this data about restoration goals tell us? First, it confirms my initial hypothesis that there are differences in approaches to ecological restoration between Europeans on one hand and Australians and North Americans on the other. Australians and North Americans have a much greater interest in the restoration of historical ecosystems than do restorationists in other parts of the world. I think that the similarities between the two areas derive from their similar histories. Both areas were fairly recently settled by Europeans. In both cases the Europeans viewed the landscape as being largely empty, despite the presence of native people, or at the least thought that the native people did not really use the land and thus did not possess it the way that Europeans possessed and used land. There was a rapid conversion of the landscape from one originally seen as wild to a highly domesticated landscape in both Australia and North America. In both places the land and resources were at first seen as being almost infinite, with nothing to limit growth and expansion. Within living memory many ecosystems were converted from seemingly wild to domesticated and in many cases these seemingly wild landscapes were completely destroyed by activities like agriculture and mineral extraction. There is both memory of what was there before and an increasingly romantic attachment to the landscape first experienced by the pioneers. But I think another critical link between Australia and North America is that both evolved from British colonies and the shared legacy of that British heritage – common language, original basis for their legal systems, attitudes toward land use and common literature (e.g., Wordsworth's poetry is well known in both areas and helps inform attitudes toward nature) – has played a large role in forging common attitudes toward restoration. I don't discount the still heavy French influence in Quebec, but it is the British heritage that is shared and separates Australia and North America from another recently settled (by Europeans) region – Latin America – which I will return to in a moment.

Restoration in the European context is largely about the preservation of biodiversity and also the enhancement of ecosystem functions and services. Ruiz-Jaen and Aide (2005) found that almost 60 percent of all restoration projects were carried out in response to a legal requirement for restoration. Because European laws and frameworks for environmental management such as Natura 2000 (Cliquet *et al.* 2009) place a high value on biodiversity, it is natural that European restoration should be primarily concerned with biodiversity. Europe has very few examples of ecosystems that North Americans and Australians would consider wild – mostly just a few forests such as in far northern Scandinavia and Białowieża in Poland. But Europe is rich in semi-natural agricultural landscapes that often contain high levels of biodiversity (Bakker and Berendse 1999). Thus Europe is the region where restorationists are most likely to restore agricultural lands to agricultural status – if that agricultural status is low-intensity, traditional agriculture that supports much biodiversity. In many ways the restoration of semi-natural pastoral landscapes is also a form of romanticism and nostalgia similar to the desire for "wild" lands in Australia and North America. Many of these semi-natural agricultural lands were farmed using traditional methods until just after World War II, when more intensive, industrial agriculture first began to be practiced on a large scale in Europe. Again, within living memory those landscapes were lost and the desire to restore them is a desire to return to a not-too-distant past even though the goal for restoration is often discussed in terms of biodiversity.

I found far fewer studies from the other major geographical regions, so it is hard to draw firm conclusions about attitudes toward restoration there. Africa, where I only found reports on 13 restoration projects, most of them in South Africa, is especially difficult to evaluate. But all of those other areas – Africa, Asia, Latin America and Oceania – appear to be more similar to Europe in terms of attitudes toward and goals for ecological restoration than they are to Australia and North America. All of them place a fairly high value on the restoration of biodiversity, ecosystem functions and ecosystem services. Asian restorations in particular seem to be largely concerned with the restoration of ecosystem function and services. The descriptions of restoration projects in China are especially likely to discuss the restoration in terms of direct practical benefits accrued via the restoration of ecosystem services. Africa and Asia, like Europe, are areas with a long history of habitation by the same groups of people, people who have used the same land-management techniques for centuries. That long history of continuous human presence and land use must be especially important in informing attitudes toward ecological restoration. If the land hasn't undergone recent radical changes in ownership or use patterns then the most important things to restore would almost certainly be lost biodiversity and ecosystem services.

Case study: large-scale ecological restoration projects in China

China is home to some of the largest and most ambitious ecological restoration projects in the world. It is unfortunate that they are not well known in the West because with their large scale and rapid deployment they have the potential to be important not just at the local or regional level, but at the global level too. There is also much that restorationists can learn about originating and implementing such wide-spread and ambitious programs, but how much can be learned will depend upon how completely the restoration methods and plans are reported and the quality of the data published about the projects.

The Chinese government undertook these large restoration projects because it recognized that following the establishment of the People's Republic of China, rapid population growth coupled with increased demand for natural resources such as wood products, agricultural production and industrialization resulted in severe environmental degradation in many parts of China. From 1949 until the late 1990s there was excessive forest clearing and deforestation, severe overgrazing and conversion of forested and grazing land to arable crop production, which led to huge losses of forests, massive problems with soil erosion and degraded grasslands no longer capable of supporting grazing animals or the rural populations that depended on those animals.

Thus, beginning in the late 1990s, the Chinese government, in particular the State Forestry Agency (SFA), began several ambitious restoration programs in order to stop further environmental degradation and to correct existing problems as completely as possible. From the beginning these programs were focused on restoring ecological function and ecosystem services. Their initial desire was to protect land that still existed in good ecological condition and to restore ecological structure and function (especially primary production and nutrient cycling) and ecosystem services (such as erosion control, flood control and socio-economic benefits of grazing agriculture and forestry) to degraded lands (Yin and Yin 2010; Yin *et al.* 2010). While restoration in America and Australia, even large-scale restoration projects, are often focused on restoring the historical trajectory to damaged ecosystems and in Europe there is a focus on restoring and maintaining biodiversity, in China the focus was more pragmatic and centered on restoring ecosystem properties of direct benefit to humans.

In order to achieve its goals, the SFA initiated a veritable alphabet soup of programs beginning in 1998. The first large-scale project was the National Forest Protection Program (NFPP). The goal for the NFPP was to conserve about 90 million hectares of remaining good, natural forest and to restore an additional 8.7 million hectares of damaged forests.

In 1999 the Sloping Land Conversion Program (SLCP – also sometimes referred to as the "Grain to Green" program) was established to reduce severe soil erosion. The goal was to remove land prone to soil erosion or already damaged by severe erosion from agricultural production by both aforestation and restoring lost grasslands. The initial goal of the SLCP was to retire 14.6 million hectares from agriculture and convert it to other land uses (mainly grassland), and to restore and plant forest on another 17.3 million hectares between 2001 and 2010. The ultimate goal of the SLCP is to have effective erosion control on 86.67 million hectares and to eliminate desertification on 102.67 million hectares.

The Desertification Combating Program around Beijing and Tianjing (DCBT) was established in 2001 in order to eliminate the severe duststorms sweeping across the Beijing–Tianjing area due to overgrazing and desertification. Plans called for the DCBT to restore 10.13 million hectares to grassland.

The Wildlife Conservation and Nature Preserves Program (WCNR – also known as the Great Green Wall) was set up in order to develop a band of wildlife habitat 400–700 km wide, extending across 4,500 km parallel to the Great Wall of China in order to protect and enhance biodiversity, promote wetland restoration and encourage ecotourism.

The SFA also has other restoration programs, as do other government agencies, but these are the four largest programs. It is estimated that the Chinese government spent $70 billion on these projects from 2001–2010 (Yin and Yin 2010; Yin *et al.* 2010). By way of comparison, the somewhat similar Conservation Reserve Program (CRP) in the United States was designed to take marginal agricultural land out of production, prop up crop prices in the United States, greatly reduce soil erosion and provide wildlife habitat. The CRP is usually considered to be a highly successful program and as of 2010 there are about 12.67 million hectares enrolled in it (USDA 2011).

Most papers published about the Chinese ecological restoration programs report great successes (Li *et al.* 2008; Zhou *et al.* 2008; Xu *et al.* 2010). In Guangdong Province, the area of land covered by forest has almost doubled from 26.2 percent of the land to 50.1 percent of the land from 1979 to 1998. The increase in forest area has resulted in carbon sequestration equal to 22.5 percent of the total CO_2 released in Guangdong Province during that time (Peng *et al.* 2009). Already, 1.25 million hectares of grassland have been restored in Northern Tibet, with a corresponding 40 percent increase in grass production and thus better support for livestock grazing (Gao *et al.* 2009).

The successes and benefits claimed for the Chinese restoration programs are extremely impressive, but some experts have suggested we need to be cautious in interpreting those reports. They are concerned that most of the available data has been collected by government agencies and there is no independent verification of the extent and impact of the restoration programs. They are worried that the restoration programs were implemented quickly, with an emphasis on quantity not quality and that the lessons learned early in the restoration process were not applied to later restorations.

The SFA is also focused on planting as many trees as quickly as possible and less focused on restoring grasslands or using natural re-vegetation methods. The SFA frequently uses non-native trees and trees that may be inappropriate for the ecological situation, such as planting water-consuming poplars at arid sites. There is also some indication that many farms and farmers were involuntarily entered into the programs (Yin and Yin 2010; Yin *et al.* 2010).

Clearly, a strongly centralized government such as the Chinese government allows large-scale planning and rapid implementation of big restoration projects. Trying to plan regional restoration projects with large, even continental and global benefits can be difficult in countries with a strong tradition of independent private land owners, such as the United States, or in a multi-nation setting like the European Union. Such projects are increasingly important as we deal with global environmental change, but the Chinese experience tells us that an open process with independent verification of achievements is vital in order for scientists and practitioners to learn as much as possible about the challenges and successes when working at such huge regional and national levels. It is also vital that individual stakeholders are consulted about restoration projects and free to opt in or not based on their own understanding of project goals.

Latin America appears to be an outlier here. Like North America and Australia, it was recently settled by Europeans. But there was something about the history of settlement and domination that has resulted in a different approach to restoration in Latin America than in North America and Australia. I suspect it was a combination of the development of a mestizo or mixed culture in Latin America, a development that was frowned upon and largely rejected in the English-speaking colonies, and the fact that Latin America attracted far fewer European immigrants and settlers than did North America.

The development of a mestizo culture preserved traditional land uses in Latin America in a way that somewhat mirrors the preservation of traditional practices in Asia and Africa so that restoration appears to be progressing along similar paths in those areas. Caboclos and caicaras, two groups of mestizo people living in parts of rural Brazil, use a combination of Native American and European practices in a way that helps maintain traditional land uses (Berkes *et al.* 2000). Another factor that may be important is that in the British-dominated colonies in North America and Australia, the British governments imported the British system of private land ownership in which there were many small land owners of European origin who had free rein to develop the land as they saw fit. In Latin America, the Spanish and Portuguese colonists imported a feudal system in which a few people owned vast tracts of land and relied on native laborers to develop and maintain the land. Often, the native laborers continued using traditional methods as they worked the land, thus traditional management methods survived in some form in many Latin American ecosystems.

Many restorationists talk about the need to involve stakeholders in restoration projects. Therefore I asked whether we are doing a good job of including stakeholders in the planning and performance of restoration projects. Unfortunately the answer is no, we are not doing well on that front (Figure 6.7). Overall only 7.7 percent of all restoration projects I surveyed reported stakeholder involvement. There are significant differences among regions with respect to stakeholder involvement ($G=22.66$, critical $\chi^2_{0.05(6)}=12.592$, $p<0.001$). Restorationists in Oceania are doing a much

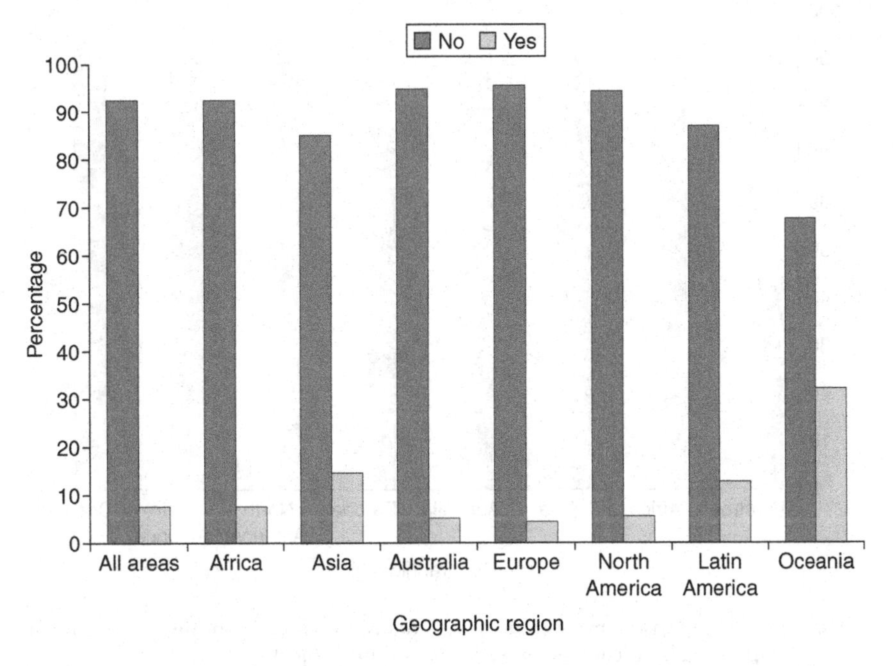

Figure 6.7 A comparison of stakeholder input into restoration projects by region.

Notes

There was a significant difference in stakeholder input among the regions ($G=22.66$, critical $\chi^2_{0.05(6)}=12.592$, $p<0.001$). Data for all areas show the combined total for all the regions and are included to show how individual regions differed from the overall group. Please note that the log-likelihood ratio was performed on count data but the graphs show percentages to allow more equitable visual comparison among geographic areas. Please also note that the data for all areas were not included in the statistical tests so as not to test the same data twice.

better job of including stakeholders (32.3 percent of projects) than are restorationists in other areas. I do not know why that is, but elsewhere restorationists must stop just talking about involving stakeholders and must actually begin to include them.

Similarly, many restorationists have discussed the need to emphasize the economic value of ecological restoration. James Aronson in particular (Aronson *et al.* 2007, 2010) thinks that we must make clear arguments about the economic benefits of restoration in order for ecological restoration to fulfill its promise as potentially the most important response to global environmental change in the twenty-first century. Do published reports of restoration projects document the economic value of ecological restoration? Unfortunately, few papers (6.3 percent overall) actually report any information about the economic benefits of restoration (Figure 6.8). There were no significant differences among geographic regions with respect to providing economic information (G=4.4, critical $\chi^2_{0.05(6)}=12.592$). Restorationists all

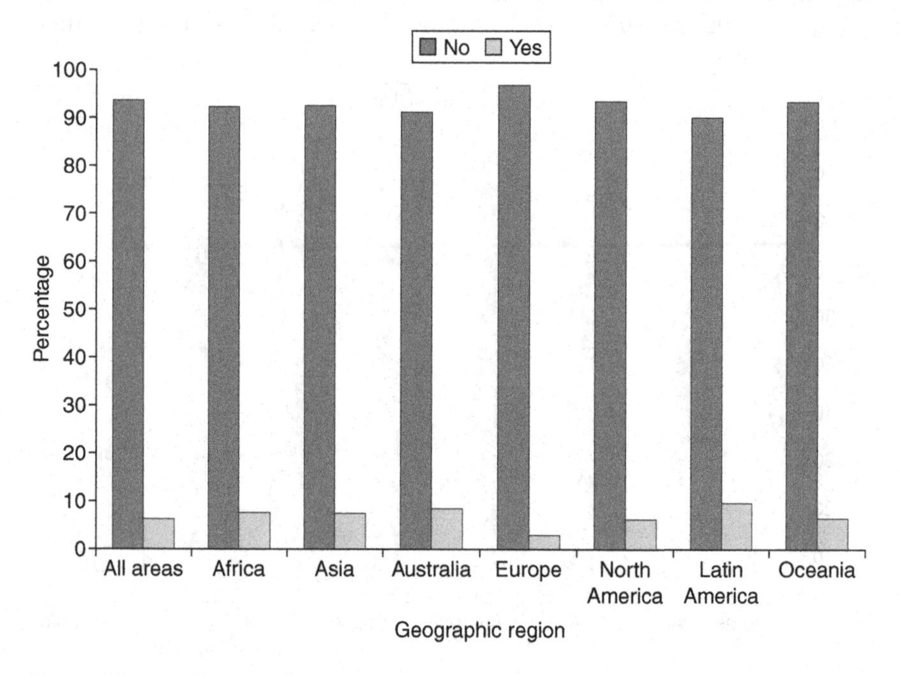

Figure 6.8 A comparison by region of the inclusion of an economic valuation in papers about each ecological restoration project.

Notes

There were no significant differences among regions with respect to including economic valuation in the published paper (G=4.4, critical $\chi^2_{0.05(6)}=12.592$). Data for all areas show the combined total for all the regions and are included to show how individual regions differed from the overall group. Please note that the log-likelihood ratio was performed on count data but the graphs show percentages to allow more equitable visual comparison among geographic areas. Please also note that the data for all areas were not included in the statistical tests so as not to test the same data twice.

across the world need to make more of an effort to calculate the economic benefits of restoration and then report those benefits to the general public.

Web-based survey of restorationists

As another way to assess restorationists' attitudes toward ecological restoration, I attempted a web-based survey of restorationists located around the world. I say attempted because in the end the survey was not completed by as many people as I hoped to reach.

Invitations to participate in the survey were included in the SER newsletter sent to all members of the SER and others (currently sent to about 2,000 people worldwide) and also in the RESTORE e-bulletin, which is sent to almost all members of the SER and other interested people (currently sent to about 1,500 people worldwide – both figures were given to me by Levi Wickwire, SER Project Coordinator, personal communication, March 31, 2011). There is much overlap between the SER newsletter and RESTORE mailing lists. Thus the survey invitations reached a total of at least 2,000 people. The survey was set up on Survey Monkey and was open from December 1, 2010 until March 8, 2011. Invitations to take part were sent in every issue of the RESTORE e-bulletin published during that time period (RESTORE is sent out weekly). The SER newsletter was sent twice when the survey was open. At the end of that time I had received 81 responses to the survey. This was far fewer than I was hoping to collect. Unfortunately there was little geographical diversity among the respondents – 70 were from North America, 3 from Australia, 2 from Europe, 2 from Latin America, 2 from Oceania and 2 declined to answer the question about geographical location. Because the geographical representation was so heavily dominated by North Americans, I don't think there is any way to make meaningful comparisons among geographical regions. And given that the sample size is smaller than I had hoped for, I also think that statistical tests of the data would be of limited value. So I have simply compiled the responses to allow for verbal comparison and discussion of them.

While 81 responses are fewer than I was hoping for, I don't think that it means that there is nothing to be learned from the survey. In fact Moore and colleagues (2009) published the results of a survey of eight especially selected experts in the field of ecology asking them to comment on how the science of ecology informs environmental management. Such a small sample (they had originally hoped to survey 20 experts) introduces the possibility of the responses being somewhat idiosyncratic and not entirely representative of the field (certainly there are more than 20 widely recognized experts in the field of ecology). I collected a random sample of 81 people working in ecological restoration, which while small, is probably no more idiosyncratic than the group Moore and colleagues surveyed. Along with being predominately North American, my respondents are also predominately male (64.5 percent). They may be youngish or at least not grizzled old-timers in the field – almost 30 percent had worked in restoration for 6–10 years, almost half had worked in restoration for 10 years or less, and over 60 percent had worked in restoration for 15 years or less (Table 6.1). Twenty-nine percent described themselves as either a restoration ecologist or ecological restoration practitioner (14.5 percent each). The next largest occupational/restorationist groupings were research scientist and natural resource manager (13.2 percent each) (Table 6.1). The survey group is engaged with the field as either professionals or active practitioners. The respondents may be more internet savvy or internet friendly than the average restorationist given their willingness to fill out the online survey.

So what do a youngish, active, internet-friendly group of restorationists think about the current state of ecological restoration? To answer that question I presented the respondents with a series of statements and asked them to tell me whether they strongly disagreed, disagreed, were neutral, agreed or strongly agreed with the statement. I also gave the respondents the opportunity to provide written comments about every question and asked them some ranking and open-ended questions, which I will discuss later.

First I presented them with the standard *SER Primer* (2004) definition of ecological restoration ("Ecological restoration is the process of assisting the recovery of an ecosystem that has been degraded, damaged or destroyed") and the statement "The SER definition matches my own working definition of ecological restoration." The vast majority of the respondents (92.5 percent) either agreed or strongly agreed with that statement (Table 6.2). Although there was widespread agreement with the standard SER definition, several respondents commented that the definition is quite general and broad.

I then presented the respondents with an expanded definition also taken from the *SER Primer* (2004):

> Ecological restoration is an intentional activity that initiates or accelerates the recovery of an ecosystem with respect to its health, integrity, and sustainability. Frequently, the ecosystem that requires restoration has been degraded, damaged, transformed or entirely destroyed as the direct or indirect result of human activities. In some cases, these impacts to ecosystems have been caused or aggravated by natural agencies such as wildfire, floods, storms, or volcanic eruption, to the point at which the ecosystem cannot recover its predisturbance state or its historic developmental trajectory. Restoration attempts to return an ecosystem to its historic trajectory.

I concluded with the statement "With the additional material, the SER definition describes my approach to ecological restoration." This expanded definition proved to be somewhat less in line with the thinking of my respondents – 70.3 percent either agreed or strongly agreed with the statement (Table 6.2). Many respondents provided written comments with

Table 6.1 Basic descriptive data about the internet survey respondents. The data is reported as percentages of the respondents who fit particular categories

Number of years working in ecological restoration	*Percentage*	*Profession/work in ecological restoration*	*Percentage*	*Source of knowledge about restoration methods and practices*	*Percentage*
1 year or less	3.9	Academic	6.6	Journals, books	19.7
2–5 years	15.6	Research scientist	13.2	Other experts	18.3
6–10 years	29.9	Restoration ecologist	14.5	Personal experience and work	16.9
11–15 years	13	Environmental consultant	7.9	Education	10.6
16–20 years	11.7	Ecological restoration practitioner	14.5	Conferences	8.2
21–25 years	6.5	Natural resource manager	13.2	Internet, web sources	6.9
26–30 years	10.4	Landscape architect	5.3	Society for Ecological Restoration	6.9
31–35 years	9.1	Land manager	6.6	Other societies	4.6
36+ years	0	Conservationist	3.9	Workshops	4.6
		Conservation planner	1.3	Government agencies	3.2
		Policy-maker	1.3		
		Other	11.8		

Table 6.2 Internet survey respondents were asked to respond to the following questions by selecting one of five possible ranks to indicate their degree of agreement or disagreement with the statement

Original question	*Strongly disagree (%)*	*Disagree (%)*	*Neutral (%)*	*Agree (%)*	*Strongly agree (%)*
1 The 2004 *SER Restoration Primer* on ecological restoration defines ecological restoration thus: "Ecological restoration is the process of assisting the recovery of an ecosystem that has been degraded, damaged or destroyed." The SER definition matches my own working definition of ecological restoration.	2.5	2.5	2.5	44.4	48.1
2 The 2004 *SER Primer* goes on to say: "Ecological restoration is an intentional activity that initiates or accelerates the recovery of an ecosystem with respect to its health, integrity and sustainability. Frequently, the ecosystem that requires restoration has been degraded, damaged, transformed or entirely destroyed as the direct or indirect result of human activities. In some cases, these impacts to ecosystems have been caused or aggravated by natural agencies such as wildfire, floods, storms or volcanic eruption, to the point at which the ecosystem cannot recover its predisturbance state or its historic developmental trajectory. Restoration attempts to return an ecosystem to its historic trajectory." With the additional material, the SER definition describes my approach to ecological restoration.	3.7	12.3	13.6	40.7	29.6
3 Ecological restoration projects have a definite end-point after which no more restoration work is necessary.	25	42.5	13.8	16.3	2.5

	Statement					
4	When or if an ecological restoration project has been finished the restored ecosystem should be indistinguishable from the pre-disturbance ecosystem.	8.8	52.5	23.8	13.8	1.3
5	Ecological restoration projects should provide opportunities for stakeholders to participate in the restoration process.	0	0	22.5	38.8	38.8
6	Ecological restoration projects should encourage humans to engage with the environment throughout the entire lifetime of the restoration.	2.5	3.8	20	41.3	32.5
7	Humans have a moral obligation to repair damaged ecosystems.	3.8	1.3	12.5	40	42.5
8	Ecological restoration projects will be an important tool for adapting to the effects of projected global climate change.	2.5	1.3	21.3	57.5	17.5
9	Ecological restoration projects will be an important tool for mitigating projected global climate change via potential benefits like sequestering atmospheric carbon.	3.8	1.3	31.3	45	18.8
10	During the twenty-first century it will be important to combine ecological restoration with agricultural practices, forestry and urban and rural land management.	1.3	1.3	2.5	43	51.9

respect to the statement. One of the main points that people chose to comment on was the vagueness or difficulty of defining particular terms such as health and integrity of an ecosystem. Because such terms are hard to define, they are hard to measure, so their inclusion in the definition makes it less useful to some respondents. The other point to attract many comments was the reference to historical trajectory and pre-disturbance state. Several respondents felt that historical trajectory is a vague idea because past history can be hard to determine and also because the choice of a time in past history on which to base the trajectory of recovery is seen as subjective. Some respondents felt that in many situations we don't know the pre-disturbance state or have only a very general understanding of what existed at the site pre-disturbance, thus making the goal of matching pre-disturbance conditions subjective as well.

Using similar statement and response questions, I found that (Table 6.2):

1 67.5 percent disagree or strongly disagree that ecological restoration projects have an end-point after which restoration work is no longer necessary.
2 61.3 percent disagree or strongly disagree that when an ecological restoration project has been finished the restored ecosystem should be indistinguishable from the pre-disturbance ecosystem.
3 Most – 77.6 percent – agree or strongly agree that stakeholders should be involved in the restoration process.
4 73.8 percent agree or strongly agree that ecological restoration projects should encourage humans to engage with the environment throughout the entire lifetime of the project.
5 82.5 percent agree or strongly agree that humans have a moral obligation to repair damaged ecosystems. Several respondents commented on the difficulty of evaluating other people's morals or generalizing about moral values.
6 75 percent agree or strongly agree that ecological restoration will be an important tool for adapting to the effects of global climate change.
7 63.8 percent agree or strongly agree that ecological restoration will be an important tool for mitigating projected global climate change via benefits like sequestering atmospheric carbon.
8 A large majority – 94.9 percent – agree or strongly agree that during the twenty-first century it will be important to combine ecological restoration with agricultural practices, forestry and urban and rural land management.

I asked respondents to rank a list of nine possible reasons for engaging in ecological restoration (Table 6.3) on a scale of 1 (most important, highest priority) to 9 (least important, lowest priority). The highest-ranked reason to engage in ecological restoration, with 35.1 percent of the

Table 6.3 Internet respondents were asked: "From the list provided below, please rank the following reasons for engaging in ecological restoration from 1 (most important, highest priority) to 9 (least important, lowest priority). I want to restore to:" The table shows the percentage of respondents that ranked each reason at a particular level and also shows the overall average rank for the reason

Reason for restoration	*Rating from 1 to 9 (%)*									*Mean rating*
	1	*2*	*3*	*4*	*5*	*6*	*7*	*8*	*9*	
Restore ecosystem functions	35.1	31.2	15.6	5.2	7.8	0	0	2.6	2.6	2.48
Preserve and maintain biodiversity	27.6	39.5	13.2	10.5	2.6	0	0	2.6	3.9	2.58
Restore to its historical, pre-disturbance condition and then allow it to evolve along an ecological trajectory	14.9	8.1	14.9	23	9.5	17.6	8.1	4.1	0	4.09
Provide ecosystem services valuable to humans	5.3	3.9	17.7	18.4	15.8	17.1	17.1	3.9	1.3	4.83
Maintain social–cultural interactions between humans and the ecosystem	1.4	5.5	15.1	16.4	20.5	20.5	15.1	5.5	0	4.99
Help counter the effects of global climate change	2.6	1.3	15.8	11.8	15.8	13.2	18.4	15.8	5.3	5.61
Restore to its historical, pre-disturbance condition and maintain that condition	5.7	8.6	8.6	7.1	11.4	14.3	12.9	15.7	15.7	5.77
Other	14.3	2.9	2.9	0	8.6	5.7	8.6	8.6	48.6	6.71
Provide opportunities for economic development	2.5	3.8	0	3.8	10	8.8	12.5	32.5	26.3	7.11

Note
I have arranged the responses from highest priority to lowest priority, but in the survey they were randomly arranged. I also asked the respondents to specify "other."

respondents ranking it as the highest priority, was restoring ecosystem functions. The second highest-ranked reason for doing ecological restoration was preserving and maintaining biodiversity, with 27.6 percent ranking it as the most important reason. Restoring an ecosystem to its historical, pre-disturbance condition and then allowing it to evolve along an ecological trajectory was ranked third on the list, and restoring an ecosystem to its historical, pre-disturbance condition and maintaining that condition was ranked as the seventh most important reason for ecological restoration. The least important reason to engage in ecological restoration was providing opportunities for economic development. I allowed respondents to select "other" as a possible response and then to tell me what they referred to as other. Many specified other as esthetics – restoration as a way to return natural beauty to a site.

I also asked respondents to evaluate the principle obstacles to effective ecological restoration. I gave them a choice of seven possible obstacles and asked them to rank them from 1 (the biggest obstacle) to 7 (the smallest obstacle) (Table 6.4). Lack of money was chosen as the biggest obstacle by 44.2 percent of the respondents, which was also reported by respondents to Cabin *et al.*'s (2010) survey. The second greatest obstacle was government policies. The third highest-ranked category was "other" and I asked respondents to specify what this was. Respondents provided a wide range of possible answers to "other," some of which were categories I included as possible obstacles (a few mentioned lack of knowledge despite it being a possible category). The most frequently mentioned "other" categories were: (1) our tendency to focus on short-term projects and our desire for fast results; and (2) a lack of public support and/or public misunderstanding of ecological restoration. Interestingly, our current state of knowledge and technical problems did not rank highly among potential obstacles to effective restoration.

In an open-ended question, I asked respondents to describe where they get their knowledge and information about restoration methods and practices. I received a wide range of answers, but was able to group them into several categories (Table 6.1). The most important sources of information (in order) were: journals and books (19.2 percent), other experts (18.3 percent), personal experience (16.9 percent) and education (10.6 percent). I received several written explanations of why particular sources of information were especially valuable, and here people often expressed strong opinions. In fact, in the written comments I could see a large gap between practitioners and researchers, which was similar to the split identified by Cabin *et al.* (2010). The researchers were usually content to say they got their information from published material or from attending conferences, but several practitioners had pointed comments to make about the value of information published in journals. A couple of comments will illustrate what several had to say: "I learn from observation of nature. I avoid published recommendations based on one or two seasons of work by

Table 6.4 Internet respondents were asked: "What are the principle obstacles to effective restoration programmes? Please rank the following possible obstacles to effective ecological restoration from 1 (biggest obstacle) to 7 (smallest obstacle)." The table shows the percentage of respondents that ranked each reason at a particular level and also shows the overall average rank for the reason

Potential obstacles	*Rating from 1 to 7 (%)*							*Mean rating*
	1	*2*	*3*	*4*	*5*	*6*	*7*	
Lack of money	44.2	23.4	10.4	15.6	1.3	1.3	3.9	2.26
Government policies	13.7	16.4	30.1	9.6	17.8	8.2	4.1	3.42
Other	40.5	8.1	2.7	5.4	0	5.4	37.8	3.84
Disagreements among stakeholders	9	16.7	17.9	16.7	19.2	15.4	5.1	3.87
Our current state of knowledge	11.5	20.5	7.7	10.3	16.7	24.4	9	4.09
Lack of human resources	1.4	12.7	22.5	26.8	15.5	14.1	7	4.13
Technical problems	0	7	14.1	18.3	25.4	23.9	11.3	4.79

Note
I have arranged the responses from biggest obstacle to smallest obstacle, but in the survey they were randomly arranged. I also asked the respondents to specify "Other."

grad students"; "From practicing land managers and those actually DOING restoration, not studying or talking about it. The science is 10 years behind the practice of restoration so primary source articles are more a way to back up what managers already know."

Finally, I asked the following open-ended question: "In the box below, please briefly discuss how you think rapid environmental change due to climate change, human land use, habitat fragmentation, invasive species and other factors will influence ecological restoration in the 21st Century." As I expected with such a broad question, the answers were all over the map. Respondents expressed a large range of opinions on the future of ecological restoration in the twenty-first century. Their responses can be placed in three broad categories. First, many commented that due to rapid environmental change, a focus on restoration of historical conditions will become increasingly difficult. Indeed, many felt that a focus on historical conditions will become absolutely impossible in the future and thus is an unrealistic goal. A typical comment was: "I think that folks will realize that the idea of restoring landscapes back to historical conditions is unachievable. The trend will be to restore lands to be resilient to disturbance and to enhance biodiversity as much as possible."

Second, many wrote that a combination of factors – especially invasive species, patterns of human land use, continuing human population growth and climate change – will combine to make it more difficult to restore ecosystems in the future. The tone of these comments ranged from the matter-of-fact, a simple acceptance that our work will be more difficult in the future, to pessimism that was close to despair at the prospect (or lack thereof) of being able to accomplish anything via ecological restoration. A couple of comments help to illustrate the depth of despair felt by some restorationists: "These factors will either positively influence eco restoration by showing the growing demand for it or may overcome it and we will all be living in one big concrete jungle"; "Frankly I am not hopeful that we will be able to recognize as a species – that our impact on the planet amounts to ecocide – and change our consumption, child-bearing and resource consumption in time."

But there were also hints of optimism among the respondents. Many felt that the current rapid environmental changes will lead to a realization that ecological restoration is absolutely essential because it provides us with a way to repair damaged ecosystems. Many noted that the field will have to evolve as rapidly as the environment changes – some see this happening, others see the need for rapid change as providing opportunities for more innovation and work in the field. Some representative comments are: "It should increase the importance of ecological restoration moving forward, as we realize the link between rapid environmental change and human land use, and the potential benefits of restoration efforts"; "We will have to reevaluate our relationships and techniques that relate to biota; sometimes looking for new solutions where classic restoration techniques are not possible"; "I believe rapid environmental change will make ecological restoration less a frivolous pursuit and more an essential action for all species. It will make the field more understood and well-respected by the general public; restoration ecologists will be seen as doctors of the ailing earth rather than crazy tree-hugging hippies."

My respondents, most of whom have been working in ecological restoration for less than 15 years, are mostly representatives of the third wave of ecological restoration I discussed in Chapter 2. They come from a cohort of restorationists who came to a somewhat established field, one with traditions based on the restoration to past historical conditions and also the recognition that some lands are so damaged that any kind of restoration is better than allowing the status quo to persist. And what they have seen in the past few years is increasing evidence that the entire global environment is changing more and more rapidly. Sometimes it seems like the environment changes for the worse from day to day.

We have seen natural disasters that are exacerbated by human use of the environment – Hurricane Katrina, which swept across and destroyed cities built on floodplains and the recent tsunami caused by the magnitude 9.0 Tohoku earthquake in Japan. Both caused enormous amounts of damage. It is still not clear how severe the effects of radiation leaking from damaged nuclear reactors at Fukushima will be. But it is clear that ecological restoration will have to be part of the response to such disasters. And so the third wave of restorationists are faced with an environment that seems to be increasingly susceptible to damage and difficult to restore. They have, by necessity, developed a more flexible attitude toward ecological restoration, which emphasizes the need to restore biodiversity, ecosystem function and ecosystem services without being constrained by a slavish requirement to match a past model. Reference ecosystems are still of vital importance – simply throwing species together willy-nilly is a recipe for failure, but the references must be based on both what is possible and what will provide ecological benefits to humans and entire ecosystems. My respondents were not, for the most part, interested in a restoration of historical conditions – which is a surprise given the largely North American make-up of the group. They are comfortable with the idea that restorations never end and will need constant management and human involvement. They are hopeful that ecological restoration can help solve many of the environmental problems facing us in the twenty-first century. They seem to be eager to get on with the actual work of ecological restoration and are less interested in the theoretical basis of the field.

Conclusions: differences among restorationists are real, the field is changing

The differences I first noticed when attending SER conferences are not just the result of me talking to a small, potentially unrepresentative subset of restorationists. There is a real difference between how and why North Americans and Australians approach ecological restoration and how and why the rest of the world does. North Americans and Australians are much more interested in the restoration of previously existing historical conditions than are restorationists from the rest of the world. Although Europeans have an interest in restoring recently damaged ecosystems to their previous conditions, those ecosystems are predominately low-intensity agricultural landscapes that support much biodiversity. There is nostalgia in that desire, just as there is nostalgia in the Australian and North American attempts to restore wild landscapes, but the motivation is different. Europeans desire a multi-functional landscape that includes high biodiversity, traditional landscape management, beauty and some wild areas (explored in more detail in Chapter 7). North Americans and Australians dream about returning to a landscape with little human influence (or at least little Industrial-Age human influence). African, Asian, European, Latin American and Oceanian restorationists are not synchronized in their motivations for carrying out ecological restoration, but they do in general emphasize restoration to promote biodiversity, ecosystem function and ecosystem services. Restorationists in those areas appear to be more comfortable using restoration as a tool to achieve goals that benefit humans as much as they benefit the rest of the biota.

But there is also evidence that the field of ecological restoration is changing. My survey of recently published literature found less evidence of a plant-oriented bias in restoration projects, more interest in a wide array of taxonomic groups and a growing interest in restoration at the level of entire ecosystems rather than particular organisms. I also found, via an online survey that reached a predominately North American audience, that restorationists in North America today are more interested in restoration of biodiversity and ecosystem function and services there too. Because of the time lag in publication – it is usually a couple of years after someone starts writing a paper before it appears in print – even a survey of recent literature may miss current developments on the ground.

Given the rapidity of global environmental change and the rise of novel ecosystems, restorationists will have to adopt restoration practices that are focused on biodiversity, ecosystem function and services first in the twenty-first century, honoring the historical past for a site second. Restoration without any recognition of the past is not restoration, but we need to be forward-thinking as we use ecological restoration to meet the challenges of the coming years. Human population growth and needs will require us to develop functional restored landscapes that benefit all of the biota, not just humans and not just wild species. Now our task is to imagine how this renewed restoration will happen and how it can be supported.

7 Renewed restoration

Building a middle path toward a restored earth

In this chapter I will:

- explain new approaches to restoration necessary in the new millennium;
- outline the form renewed restoration should take;
- discuss how restoration can help rebuild the social–cultural connection to the natural world;
- discuss the challenges inherent with a broadly inclusive restoration program – how can we adequately incorporate all stakeholders in restoration?;
- ask whether pure or applied science is better suited to serve the needs of renewed restoration;
- show how the creative arts can help us better understand the process of restoration;
- examine the role of tradition and celebration in creating and maintaining the restoration community.

Introduction: new approaches to restoration needed in the new millennium

In some ways it already feels passé to talk about the new millennium – after all, the celebrations to usher in its arrival took place over 11 years ago. My daughters have no memory of the past millennium even though the oldest was born before it ended. But given that a millennium lasts 1,000 years, we are still in the early days of the new one. Even the twenty-first century is still young. But time races forward ever faster – and not just because I'm getting older myself and I notice new years seem to arrive with greater and greater rapidity.

My great-grandmother Rankin, who I knew well, was born in 1870 and lived to be 103. I used to marvel at the amazing technological changes she observed. She grew up on a farm in rural Illinois at a time when farm labor was all accomplished by muscle power, whether supplied by humans or horses. Indoor plumbing and electricity were unknown to her. She must have been in her thirties before she saw her first automobile or airplane. And yet she lived long enough to watch astronauts walking on the moon on her television. I once thought that never again would a single person's century-long life witness such profound change.

But now I am not so certain about that. My students are shocked when I tell them that I typed my Master's thesis on an electric typewriter and that I once wrote programs on punch cards and stored data on rolls of magnetic tape. I still have those magnetic tapes somewhere, along with a box full of floppy disks, and I doubt any of them are still readable. My oldest daughter once asked what my favorite computer game was when I was a child and could not believe that we did not have computer games or home computers when I was a child. My students and children have seen and even used landline phones, so they know there was once a time when cell phones did not exist, but they can't imagine their lives without their cell phones and instant access to the entire internet. Now I think that there is a good chance that if I live to be 100, I will see changes more profound than great-grandmother Rankin. If my daughters live 100 years, they will almost certainly see changes even greater than I did. But what will those changes be and will they be good changes?

Ecological restoration today takes place in a world wracked by the dizzying pace of constant change. Our technology and our demand for its products – mobile phones, MP3 music players, digital cameras – also race forward like a speeding locomotive (an already archaic term). As the global population grows, the demands we place on the earth and its ecosystems grow larger with the same headlong rush. But, unfortunately, ecosystems and their species do not grow, recover and repair themselves at our ever-increasing human-imposed technological tempo. The natural world continues as it always has, moving at ecological pace, its dynamic cycles working in evolutionary time. We can damage the world ever more rapidly at larger and larger scales, but on its own the world recovers at the same pace it has ever since complex life evolved hundreds of millions of years ago. The geological record tells us that it takes about ten million years for biodiversity to recover from a mass extinction, and many people fear that we are causing a mass extinction event right now (Barnosky *et al.* 2011). In the lifetime of the earth, ten million years isn't very long, but in the lifetime of a human or even our entire species, ten million years is an unimaginably long time.

In the twenty-first century it is increasingly important that ecological restoration helps to close the gap between the human sprint towards new and more and the earth's slow and steady recovery from our damages. The twenty-first century requires renewed ecological restoration that traces a middle path between many different styles of and competing visions of what ecological restoration can and should be. In previous chapters I have identified the primary challenges to ecological restoration as rapid climate change and the rise of novel ecosystems. But for restoration to truly be successful and fulfill a vital role in repairing the damaged relationship between humans and the rest of the earth, it will also have to address the increasingly huge human footprint and global inequities in the use of resources. Ecological restoration will have to be truly global to help solve such huge problems, but as we shall see, there is good evidence that restoration is easiest to achieve at small, local scales. In *Ecological Restoration: A Global Challenge*, Francisco Comin (2010) wrote that we need to expand our environmental mantra beyond the traditional three Rs – Reduce, Reuse and Recycle – to include a fourth R – Restore. He thinks we even need to move beyond four Rs to add a fifth – Rethink our use of resources. Comin is right – Restoration will have to become as well known as the three Rs and an integral part of our Rethinking our relationship to the world.

Renewed restoration

The key issue that must be addressed in renewed restoration is combining the restoration of damaged ecosystems with the restoration of the human–nature connection. For many years commentators on ecological restoration have recognized that restoration of ecosystems and the human–nature connection must be two parts of a whole if restoration is to fulfill its promise and potential as a force revitalizing both people and the environment (Light 2000; Higgs 2003; Jordan 2003). Unfortunately, in practice, restoration professionals usually focus on just one part of the whole (almost always ecosystem restoration) and don't incorporate both halves in either the design or implementation of restoration projects. As a result there is a great amount of emphasis on the scientific/technical side of restoration practice, but only a little emphasis on the human side of restoration. But I should insert a caveat here – the emphasis on the scientific and technical is readily apparent when reading the published literature on ecological restoration, when listening to presentations at restoration conferences and even in informal after-hours discussions among members of the Society for Ecological Restoration (SER). Those forums tend to be dominated by professionals working in restorations – professionals who are almost exclusively trained in science and the technical aspects of restoration and who are most comfortable speaking about their areas of expertise using the language they have been trained to use. But there is a large community of restorationists out there who are not professionally trained, and they are a much harder group to characterize with respect to attitudes toward restoration. Lay practitioners are not as prominent in the most public forums about restoration and thus it is hard to determine why exactly they are engaged in restoration. Some, like Jay Stacy at Nachusa Grasslands and Rob Mungovan at Fowlmere, cite concerns about biodiversity and ecosystem function when asked why they engage in restoration. But it is obvious when talking with them that they derive a deep sense of personal satisfaction, fulfillment and connection to nature via their restoration work. Even if they adopt the language of professionals when talking about restoration, they possess a deep passion that can be seen by the fire in their eyes and their eagerness to share their work with others.

Here, at the beginning of the twenty-first century, there is an increasing emphasis on large-scale, highly technical restoration projects as we attempt to address global environmental problems (Higgs 2010). This will sound contradictory because later in this chapter I will argue that large-scale projects are absolutely necessary for accomplishing renewed restoration but, while large-scale projects are necessary for renewed restoration, they are not sufficient to accomplish renewed restoration. Renewed restoration – making Restoration the fourth R and Rethinking the fifth R (and perhaps Renewal the sixth R) – depends on more than just grand scale and good technique; it is also absolutely dependent on a restored human connection to the natural world. And restoring the human connection to nature happens individual by individual, each working on some small piece of land important to them.

It is easy to understand where this focus on large-scale, technical restoration originates. We look at the world around us and see that environmental problems start with local actions, but the combined effects are global. Only global, coordinated planning and implementation can solve such huge problems. The environmental conferences at Rio, Kyoto and Copenhagen have created global frameworks for tackling environmental degradation. The common language spoken at those conferences and by large national and multi-national agencies is technical, scientific, precise and as objective as possible. There are advantages to that approach. Scientists, engineers and planners can all speak and understand that language and thus it is possible for a huge, multi-national group to understand the scope of the issues facing us, even if they don't always agree on how to solve them or the urgency with which we need to act.

Many in the restoration community are increasingly interested in ecological restoration as a tool to restore ecosystem services and in measuring direct economic benefits of restoration (Aronson *et al.* 2007). Many are especially interested in discussing restoration in terms of the TEEB project (The Economics of Ecosystems and Biodiversity, www.teebweb.org, initiated by the UN Development Programme in 2007). The respondents to my online survey cited a desire to restore ecosystem functions and biodiversity as the main reasons for engaging in restoration.

The conclusions of economic analyses of the value of restoration are stark and sobering. Robert Costanza made economic comparisons between a business as usual (BAU) earth, in which we continue to use resources at our current rate and a Restored (RE) earth, in which we drastically change how we treat resources and the earth. He envisions an RE earth in which 15 percent of the terrestrial and 30 percent of the marine environments are restored and protected from development. Adaptive management strategies are used for the rest of the world to enhance ecosystem services and to use resources sustainably. The BAU earth is predicted to suffer a 10 percent decline in the value of ecosystem services from 2000 to 2050 – a loss of about $3.5 trillion worth of services per year. Costanza thinks his estimates are conservative because in his models he did not include any ecosystem suffering complete collapse, which is a possibility given the current rate of climate change. In contrast, the RE earth is predicted to experience a 7 percent increase in the value of ecosystem services from 2000 to 2050 and would return ecosystems to 87 percent of their value in 1900, back when the world was more agricultural and wild and before rapid population growth and technological development (Costanza 2010). As I said, such calculations are sobering, but they don't get me fired up to go out and restore an ecosystem. I don't get up in the morning thinking "Okay, today I'm off to Green Oaks to do some savanna restoration that will contribute a small portion of the total 7 percent increase in global ecosystem services." And I doubt that many other restorationists go off to the field thinking about economic improvements either.

For many years Eric Higgs has worried about a split in the restoration community between the professionals and the community of lay restorationists. He once identified the gap between the groups as a "two-culture" problem in which professionals and lay restorationists are separated by language and goals (Higgs 2005). He thinks the gap has narrowed recently (Higgs 2010), but my own survey and conversations leave me less sanguine about a narrowing of the rift. Robert Cabin (2011) thinks the gap between professionals and practitioners is as great as ever – and perhaps even getting worse. Cabin, in particular, likes ecological restoration that is experimental in the sense that the outcome of the process is unknown but not in the sense of many small replicates and controls (the usual experimental method for restoration scientists) as those requirements slow us down when our main goal is to restore an ecosystem and also may turn off many practitioners.

Higgs breaks restoration practice down into two categories – technological restoration, which is focused on science and the restoration of natural capital (following the economic models of Costanza and others), and focal restoration, which he describes as ensuring restoration has a "social embrace" which "enlivens social diversity, participation, economic livelihoods, political integrity, and well-considered values" (Higgs 2010, p. 91). Higgs' fear is that a focus on ecosystem services and TEEB will tend to place a high value on technological restoration carried out by a scientific/technical cadre of professionals directed or authorized by national and multi-national agencies, laws, regulations and protocols. A growing emphasis on technological restoration will almost certainly place a high value on efficiency, uniform methods and large scale. He worries that it will not have room for the smaller-scale, individual and experiential projects typical of community-based focal restoration (Higgs 2010). I suspect he is right. I'm afraid that two things will be lost by focusing on technological restoration. The first is the individual brilliance and beauty of small projects such as Jay Stacy's prairie plantings and Rob Mungovan's chalk stream. The second is the coming together of committed restorationists working with common purpose (such as the up to 30,000 restorationists working to restore savanna ecosystems in the Chicago region on any given weekend), who form a community that transcends the restoration itself (Gobster and Hull 2000). Who would have thought that Carl Sandburg's Chicago,

> Hog Butcher for the World,
> Tool Maker, Stacker of Wheat,
> Player with Railroads and the Nation's Freight Handler;
> Stormy, husky, brawling,
> City of the Big Shoulders:

would become a green beacon, a pioneer in restoration? And yet it has, all because of an inspired group of amateur restorationists, taking time from their busy schedules, for many reasons, to come together and rebuild an ecosystem lost for almost 100 years (Stevens 1995). That is what renewed restoration is about and that is where its potential lies.

Rebuilding the social–cultural connection to the natural world

As noted in Chapter 2, ecological restoration has passed through several stages in its development and each stage has highlighted different goals for restoration projects. In the beginning of academic restoration, the projects were often focused on recreating historical ecosystems. Later, the focus shifted to using the past as a template and returning the ecosystem to its historical trajectory. More recently there has been an increased emphasis on using restoration to produce specific desirable outcomes such as increased biodiversity, renewed ecosystem function and valuable ecosystem services. We almost always try to create a particular structure (physical environment and species composition) that will produce a particular function or product. What is all too frequently common to these approaches is separation between the ecosystems – whether wild or restored – and the human communities that live nearby or within them. Renewed restoration cannot occur without recognizing how completely intertwined the human community and the wild ecosystems are now, and always have been in the past. Indeed, it is essential that renewed restoration embed that intimate, intertwining in the restoration process.

Restoration that does not reflect the past by building on the historical ecosystem and returning it to some aspect of its historical trajectory is not restoration. But in today's world, it is vital to recognize the role of human culture in shaping and maintaining the earth's ecosystems. It is more common and easier to recognize that role in Europe than in North America and Australia because the European landscape has been so thoroughly transformed by human activity that most of the continent is semi-natural or completely domesticated and has been for many centuries (Rackham 1986). Many of the most diverse and unique ecosystems in Europe are semi-natural and cannot exist without continued traditional, low-intensity human use. There is still a tendency for many North Americans to regard their continent as pristine, wild, untrammeled prior to European arrival. However, we know native peoples in North America made extensive changes to the landscape and actively managed ecosystems in order to promote the provision of desirable conditions and resources, but we often fail to fully acknowledge just how extensive such modifications were and how they continue to shape ecosystem structure and function even today (Denevan 1990; McCann 1999a, 1999b).

In many ways the European experience gives us a preview of how ecosystems in the rest of the world will be affected as world populations grow, landscapes become increasingly modified and climate change creates additional stresses for the landscape. Europe already has dense human populations and highly modified landscapes. Almost all European ecosystems have been so heavily impacted by human activity that they are best thought of as ranging from semi-natural to artificial. The only extensive areas of wild ecosystems are found in the European Arctic (Madgwick and Jones 2002). The choices and compromises that have to be made during the course of ecological restoration in Europe are likely to become the norm everywhere as we attempt to balance strategies to restore damaged ecosystems with human needs and desires.

One of Eric Katz's main criticisms of ecological restoration was to say that restoration is a big lie – it cannot recreate natural wild conditions because restoration is really human domination of nature and the imposition of human will on the landscape (Katz 1992). To a certain extent he is right about continued human domination, but ultimately it is a moot point. For better or – more frequently – for worse, humans do dominate nature and they have ever since they domesticated fire and developed tools. Ecological restoration is about changing that relationship from a lopsided, one-way use of ecosystems and resources for our benefit with no thought of the consequences to other species or future generations of humans, to a relationship of mutual interaction in which humans learn to work with ecosystems for the benefit of both. As long as we have the most rudimentary tools (and we always will) we have the potential to dominate and destroy wild ecosystems. Restoration is about recognizing the destructiveness of our actions and reining in our behavior so that we become part of the ecosystem, not its destroyer.

How do we use restoration to affect that renewal in our relationship to the wild? One way is to preserve and reflect on our history of past land use so we remember how we got to where we are today. The restoration of semi-natural ecosystems like European chalk grasslands or the moors in Dartmoor and Exmoor are exemplary cases of restorations that preserve the history of past land use. In those ecosystems the entire purpose of the restoration is to maintain the semi-natural, human created condition due to its value both for preserving biodiversity and a landscape with cultural meaning. However, maintaining the ecological and cultural value of chalk grasslands and moors are examples of restoration that are almost too easy to support because the public is usually in favor of their restoration and has positive associations with them. It is the difficult cases that are most interesting because they tell us so much about our underlying assumptions of why we restore ecosystems and what it means to preserve cultural history and meaning.

Post-mining and post-industrial landscapes are often littered with unused, derelict and severely degraded ecosystems. Often these sites have no remnants of the previously existing ecosystem; the soil and original topography have been drastically impacted and may be completely lost; toxic chemicals may be present; and the sites are often highly eroded. They are extremely difficult to restore because of the degree of environmental damage. In some cases the best course is to do the absolute minimum of restoration – remove or isolate hazardous areas, do some regrading of the land to minimize erosion and then allow natural dispersal of species to repopulate the site (Prach *et al.* 2001a, 2001b; Tischew and Kirmer 2007). However, low-intensity restoration of such sites is controversial partly because we think we should restore the damaged ecosystem more quickly and partly because we don't like to see reminders of past poor practices and the resulting damage remain on the landscape.

The post-industrial landscape is often the result of large changes in the local economy. Industrial landscapes, whether in the Ruhr, the English midlands or Pittsburg, Pennsylvania were originally developed as industries grew up around easily extracted sources of local raw materials. Once the raw materials were exhausted or when it became less expensive to separate manufacture from the source of the raw materials, corporations moved industrial sites to areas with cheap labor and the original industrial centers were simply abandoned and left as a blight on the land. For local populations these sites can be both an eyesore and a reminder of a devastated economy and the loss of jobs and an entire way of life. Thus there is frequently a sentiment that we should completely remove those reminders of the past and start anew with something better. There are many different strategies for redeveloping derelict post-industrial land, and it usually centers on producing some kind of economic use such as new industrial and commercial sites or housing. But often the remaining local population and economy can't sustain such development (Ling *et al.* 2003). Another option is to instead implement a much broader kind of restoration in these areas that incorporates restoration of biodiversity and wild nature, as well as providing for economic uses. The ideal response is to use restoration to create a multi-functional landscape. There are many functions that can be incorporated in a post-industrial landscape. Ecological functions provide space for biodiversity, wild nature and ecosystem services. Economic functions provide room for production. Socio-cultural and historical functions allow for recreation as well as opportunities for connection with the past and maintenance of local identity. Esthetic functions encourage people to move in the landscape and experience what it has to offer (Ling *et al.* 2003).

Such a broad, multi-functional landscape forms what John Rodwell has called an ecoscape. Ecoscapes are geographical mapping units based on the geology and soils, but each also has characteristic vegetation with accompanying typical ecological processes. Ecoscapes incorporate a historical–cultural narrative of past and present land use. Rodwell advocates using ecoscapes in process-oriented landscape management (Rodwell and Skelcher 2002). The historical narrative in a particular ecoscape is almost always complex. Local people may experience regret or melancholy when viewing the remains of lost industry and careers, but they also take pride in their history as a mining or manufacturing community. Thus restoration should restore biodiversity and ecosystem function while complementing new economic development, but the restoration itself also should preserve reminders of past land use (Figures 7.1–7.2 – brickworks at the

Figure 7.1 Ecological restoration at the Forest Centre & Millennium Country Park at Marston Vale, Bedfordshire, February 2011. The brick-making industry just outside of Bedford left behind extensive clay pits and industrial waste. Much of this area has been restored to create grasslands, wetlands and open water habitat. This photograph shows a small wetland area with a larger lake in the old clay pit behind it.

Figure 7.2 A close-up of the remains of the former brick-manufacturing center that was once located at the Forest of Marston Vale.

Forest Centre & Millennium Country Park at Marston Vale). Thus, Ling and colleagues (2003) wrote that restoration of England's Dearne Valley should retain visual cues to the valley's past – old mine pits, spoil banks and chimneys – as reminders of how the valley developed and that what we see today is a new ecosystem shaped by both past human history and natural recovery.

Retaining reminders of an industrial past is likely to be controversial in restoration projects, even though it serves to inform us of the human–nature relationship and how well that relationship has (or has not) worked in the past. But even more subtle reminders of past land-use are sometimes discouraged during restoration and preservation of wild areas. William Cronon describes the process of land preservation and restoration in the Apostle Islands National Lakeshore in his seminal essay "The Riddle of the Apostle Islands" (2003). The Apostle Islands are a group of 22 islands located in Lake Superior at the tip of the Bayfield Peninsula in Wisconsin. I visit them every summer and have been fascinated by their natural and human history ever since I first saw them. The Apostle Islands National Lakeshore was created in 1970 in order to protect the islands' scenic, scientific, historical, geological and archeological features for the education, inspiration and enjoyment of the general public. About 80 percent of the National Lakeshore was designated the Gaylord Nelson

Wilderness in 2004, and when people talk about the Apostle Islands it is frequently in terms of the islands being a wilderness, practically untouched by humans, undamaged and pristine.

However, as Cronon points out, the Apostle Islands have a long history of human habitation and exploitation. Native Americans, most recently the Ojibwe People, have lived there for centuries, hunting, fishing and farming. French fur trappers arrived in the seventeenth century and established a thriving trading post there. In the nineteenth century the islands were home to commercial fishing, logging and brownstone mining operations. Almost all of the islands were clear cut at that time and farmed after that. Late in the nineteenth century a tourist industry developed on the islands and tourism continues to this day. The US National Park Service frequently eliminates all traces of past human use from the lands within the National Park system. Cronon argues that in the Apostle Islands (and by extension all National Parks) we should restore wild nature as much as possible, but we also need to preserve evidence of past human use as well – whether that was use by native people or by later farmers, loggers, fishers and miners. If we lose all markers of past human activity on the landscape we also lose the memory of how that land and its current ecology developed and the historical–cultural meaning of the land (Cronon 2003). This is a controversial claim to make in the United States, but it is a claim worth making and fighting for. A renewed restoration will be most successful when we recognize the full range of past land use and the value that land has held for previous and current generations.

Case study: Landscape Park Duisburg-Nord – the restoration and preservation of a post-industrial landscape

One of the most intriguing examples of restoration preserving human cultural history occurred with the restoration of *Industrinatureal* in Germany's Ruhr district. Here old, abandoned industrial sites were made (somewhat) safe and allowed to go through a natural succession/recovery process. In many ways the principle employed is a principle of profound humility – because we don't know exactly how best to restore this land at this time, we will save it in its current condition until we do know how to treat it.

In the Ruhr the collapse of coal, iron and steel industries in the 1970s and 1980s left behind a bizarre, devastated landscape scarred with slag heaps, subsiding land forming boggy marshes and soil contaminated with organic chemicals and heavy metals. What could be done to restore such land? One obvious choice would be to remove all vestiges of industry, grade the site and smooth the contours, cap toxic areas with clay and make extensive plantings to bring nature back to the sites. But the obvious choice was also an extremely expensive choice, coming at a time and in a place where the economy was suffering, and it was also a choice that would wipe out past cultural and ecological history, leaving behind an ecosystem unlike anything that had ever existed there before.

Another response was to take what was left behind and instead of replacing it, reimagine it and reuse it in new ways that allowed for both wild development and historical preservation. The landscape architect Peter Latz designed just such an imaginative restoration project at what would become the Landscape Park Duisburg-Nord (Latz 2001). When the creation of the landscape park began in 1991 it was a 230 ha post-industrial wasteland. Latz proposed and implemented a metamorphosis for the site in which the industrial remains were reused rather than replaced. Old blast furnaces and chimneys became sculptures and climbing structures, as well as reminders of the industrial past. The old ore bunkers became rock walls and cliff faces for sunken gardens. The most heavily contaminated sites were closed off to human traffic and in some cases areas were cleared of hazards. But in general Latz preferred not to seal the hazardous sites in deep layers of clay. He preferred to allow natural off-gassing and recovery of the toxic areas. He also preferred to allow natural re-vegetation to occur so that new growth of plants is haphazard and tells the story of which sites will best support new life and which won't.

For Latz the restoration of post-industrial sites requires a new approach that accepts and works with the site as it exists now – its current physical condition complete with altered topography and damaged soils. In planning post-industrial restorations we need to ask which areas we plan to use and therefore which need the most restoration in order to remove hazardous conditions. But Latz thinks that massive reclamation efforts – such as completely removing all old machinery and equipment and regrading the site – destroys it a second time, removing vestiges of its past and its value to people and replacing it with something sterile, uniform and ultimately uninteresting. With this new approach to restoration, the restored site becomes a place where wild nature itself creates a new ecology from what remains and we can observe that rebirth as it happens (Latz 2001).

Industrinatureal and preserving the entire cultural history of a landscape – whether in the Ruhr, Dearne Valley or the Apostle Islands – is restoration of a high order conceptually. It challenges our notions about restoration because it leaves so much evidence of past land use in place. Yet it honors our traditions of ecological restoration because it preserves the historical trajectory and allows – even encourages – wild processes to produce a re-wilding of damaged sites. I'm not sure how well the German approach would work in other countries. Landscape Park Duisburg-Nord and other similar sites have rusting, decaying industrial equipment and buildings sitting in place, with signs reminding visitors that they approach such objects at their own risk. Perhaps such a warning and the assumption of personal risk works in Germany – if you are foolish enough to climb one of those structures and get hurt, well, that is your problem. You were warned. The United States is a highly litigious nation with liability laws that place blame on whoever leaves dangerous equipment in place and accessible to visitors, so such a park may not be possible there. But it should be possible to preserve benign reminders of past land use so that restoration preserves cultural history and meaning while also re-establishing biodiversity and wild function in a once human-dominated landscape.

Renewed restoration: who is involved and how?

If renewed restoration is to combine natural and human history to create restorations of both ecological and cultural value, then a wide network of people must take part in its design and implementation. But who is involved and when and how they are involved is a complex issue that is not easily answered. When Keith Bowers, the president and founder of Biohabitats Inc., a large restoration company, was chairman of the board of the SER he wrote that successful ecological restoration depends on many people who interact in a network that is far wider than scientists, professional restorationists and even lay practitioners. The restoration network must include all the stakeholders, all who have an interest in the project and its outcome. Bowers worried that if restoration projects did not start from the human perspective, then the project would simply be a technical fix of little lasting value or impact for the people most affected and who might most benefit from the restoration (Bowers 2006). Restoration cannot just be a project for experts and interested volunteers; it has to include the entire human community in meaningful ways. Unfortunately, we have few good examples of how to develop a wide network of professional restorationists, lay practitioners and interested stakeholders in a way that results in successful, widely supported restorations.

The challenge of developing a renewed restoration in the twenty-first century is profound. We want a restoration program that produces regional- or even global-scale restorations that create linkages across boundaries. Ideally we would like to create something like Costanza's (2010) RE earth, with 15 percent of terrestrial systems restored and remaining lands carefully managed to promote sustainability. But restoration on such a grand scale leaves us with a true stumper: How do we include a wide network of stakeholders in such a huge undertaking?

Usually when we think about planning and implementing large-scale projects, we start at the top, with large government or multi-national organizations developing the design and protocol. Then the overall plan is passed down to more and more local organizations, authorities and individuals and it is at the local level that the project is actually implemented. But simply imposing a set of plans and regulations from the top has not worked well in many ecological management programs, whether they are located in Canada, the United States, China or the European Union (McClain and Lee 1996; Keulartz 2009; Yin and Yin 2010). Big projects require the input of big planning groups, but if the flow of information, regulations and requirements are all in one direction – from a large bureaucratic organization to local individuals who have to live with the results of that plan – there is a high potential for resentment to build up. Resentment almost always results in a failure to achieve the grand outcome envisioned during the initial planning process.

An excellent case in point occurred with the implementation of policies developed as part of Natura 2000. Natura 2000 was encompassed in the European Union's Habitat Directive of 1992. The Habitat Directive itself was adopted when the Netherlands was chairing the European Union. The directive was based on the Dutch Nature Policy Plan of 1990. The Habitat Directive and the Dutch Nature Policy Plan both originated from excellent intentions. Prior to their implementation, much European environmental policy had been essentially a defensive strategy of preserving whatever small pieces of wild and semi-wild land remained. The new plans were based on a proactive strategy of creating new wild and semi-wild areas (referred to as "New Nature" in the Netherlands) so that there would be an increase of area devoted to promoting biodiversity and ecosystem services. In the Netherlands the goal was to have about 21 percent of the land enrolled in a system of reserves and restored areas. Unfortunately Natura 2000 was initially implemented in a top-down manner with little consultation with local managers and stakeholders – those who would actually create, manage and live with the new reserves and restorations. The initial lack of local input resulted in considerable resistance to the plans so that in many cases it became almost impossible to fulfill Natura 2000's plans and regulations (Keulartz 2009).

Once people at the government and regulatory agencies recognized there was resistance to the plans they responded by stepping back and encouraging more participation in both the planning and implementation phases. Decisions were not made until there had been considerable public consultation and interactive governance. How much public participation actually occurs during Natura 2000 planning varies from country to country within the European Union. Some countries, such as France and Germany, seem to talk about public participation, but not much participation occurs in practice. Even when there is much public participation, it is not clear that it is always a good thing. Receiving input from many stakeholders who often have differing needs, desires and visions for what should happen is expensive and takes time. Thus the implementation phase is delayed even though in some cases quick action is required to save and restore the ecosystem. There is also evidence that increased participation may result in less land being preserved and restored, and that the resulting restorations may be of lower quality due to compromises made during the planning and implementation phases (Keulartz 2009).

Unfortunately, the initial top-down imposition of Natura 2000's regulations has resulted in some distrust of the scientists and government planners working to design and monitor the projects. In many cases conflicts about the implementation of plans have arisen between expert professional scientists and planners who possess a broad understanding of scientific and engineering principles, and stakeholders with considerable local knowledge and experience but little scientific training. Because local stakeholders developed a poor perception of scientists and their working methods during the early days of top-down regulation, they prefer to rely on local knowledge when preserving and restoring land (Keulartz 2009). I do not mean to imply that distant scientists applying general principles are always right in their recommendations or that local stakeholders applying generations of experience are wrong, I just want to point out that the conversations that need to occur can be difficult, given the past history of one-sided interactions. Usually a careful balance of scientific principles and local knowledge is necessary to create successful restoration.

I also do not want to give the impression that I think large-scale projects are always doomed to fail because it is impossible to communicate across several levels, from the multi-national, national and regional down to the local. Within Europe there is an example of another large, multi-national project that has been fairly successful at combining the large scale with the local. The European Green Belt (EGB) is the effort to create a network of conservation areas along the old Iron Curtain – the boundary that separated capitalist from socialist Europe during the Cold War. The end of the Cold War and the removal of barriers in the early 1990s created the opportunity to put the EGB together. Because it is such a large network – stretching across the continent from the Barents Sea to the Adriatic and Black Seas – touching so many countries, its development has been a complex process. The EGB consists of three related but slightly different entities: (1) specific areas of land with particular ecological conditions; (2) a new type of environmental governance that combines those areas and manages them for ecological and social–cultural purposes; and (3) a group of government agencies, NGOs and academics who design and put the EGB into practice (Kortelainen 2010).

The Cold War borders were closely guarded and often quite wide on the socialist side. The border areas were not used for economic or other human purposes and thus they were left alone and gradually developed into areas of ecological importance. As early as the 1970s people recognized that the border areas had become wild and semi-wild lands that supported many rare species. The timing of the end of the Cold War was fortuitous because it coincided with a period of increasing awareness of environmental problems and the need to preserve and restore wild lands. Thus it ended at a time when there was space to think about using the border areas for conservation and other green projects. Many visionary people, such as Professor Michael Succow at the University of Greifswald, worked hard to convince governments that these border areas represented an unprecedented opportunity to preserve land, biodiversity and restore ecosystem function across a broad swath of Europe. Succow won a Right Livelihood Award in 1997 for his efforts to establish a network of reserves in the former German Democratic Republic. Development of the EGB has worked well because from the beginning the planners recognized the need to work at the multi-national level and at the local level. Thus, planning has incorporated communication and negotiations across many levels, resulting in a governance structure that receives input from all sectors involved in the EGB. The EGB is designed to foster connections and allow mobility between reserve areas both in terms of ecological connections for wild species and socio-cultural connections for people living in the EGB zone. Part of the original impetus for the EGB was to remove the stigma of the Iron Curtain from the former border area and to reduce its importance as a divider between nations and people. Ironically, the EGB exists because of those boundaries and depends on the boundaries being maintained because the borders are where the wild green space occurs (Kortelainen 2010).

One question that arises when trying to implement large-scale plans while including input from all stakeholders is how to make the process both as democratic and as transparent as possible. If there are differences in goals and perceived needs between national- and multi-national-level planners and local stakeholders, whose opinion should prevail? If a national majority favors conservation of a particular area but the local population is against it, how should we proceed? These are not trivial issues and they do not have universally applicable answers. The final decision in such matters will almost certainly differ from one situation to another. Frequently there is a rural–urban divide (which may also reflect divisions in income, social class, etc.) when it comes to differences in opinion about restoration goals. A Norwegian plan to reintroduce predators such as bears and lynx to northern rural forests led to considerable conflict between rural people who had to live near the restored predators and therefore were against their reintroduction and urban people who favored their reintroduction but would not come into much contact with the predators (Keulartz 2009). Similar conflicts arose in the United States when wolves were reintroduced to Yellowstone National Park. If the interests of a national majority outweigh those of a local population, then some kind of compensation will have to be made available to local people negatively affected by the restoration. If it is at all possible, a governance process that involves careful consultation and communication should be used to minimize conflict and hopefully reach a compromise that is both ecologically and socially viable. In cases where compromise is not possible, the Wisdom of Solomon will be necessary to reach a final decision.

In many parts of the world – the Americas, Australia, Africa, Oceania, much of Asia and even parts of Europe – who counts as local and possessing local knowledge can be a thorny issue. The restoration community increasingly recognizes that Western scientific approaches are not the only way to restore damaged ecosystems. Local people over many generations developed traditional ecological knowledge with insights to ecological processes that frequently reveal a deep understanding of connections and processes that have escaped Western scientists and are every bit as detailed and sophisticated as scientific knowledge. Native people with extensive local knowledge have forcefully pointed out the subtlety and deep understanding embodied in traditional ecological knowledge. I can still feel the chill that ran down my spine when listening to David Claudie of the Chuulangun Aboriginal Corporation deliver a keynote address at the 2009 SER meeting in Perth, Australia. He looked out at the overwhelmingly white audience and in a deep, powerful voice told us, "You people don't know anything" about managing Australian ecosystems. As he spoke I could feel the oxygen being sucked out of the room. No one, myself included, had the temerity to say "We know things; we just know different things than you do." It was our turn to be chastened and we took it.

But what counts as traditional ecological knowledge (TEK) and who possesses it? In many ways science and TEK are not so different to each other. Both science and TEK originate from repeated observations of natural phenomena. TEK can be defined as "a cumulative body of knowledge, practice and belief, evolving by adaptive processes and handed down through generations by cultural transmission.... TEK is an attribute of societies with historical continuity in resource use practice" (Berkes *et al.* 2000, p. 1252). Sometimes science is presented as originating in or at least privileging abstract theory while TEK is more concrete and historical, but that does not differentiate TEK from purely empirical science. A more important difference is that TEK incorporates moral, ethical and spiritual worldviews and places a high value on gradually evolved human wisdom (Ford and Martinez 2000). In contrast, Western science claims to be objective and as such is not concerned with moral values and ethics, although how science is applied has clear moral and ethical considerations. But the critical similarity is that both science and TEK are constantly evolving – as new knowledge is acquired, new practices may develop. Science is primarily quantitative, while TEK is more qualitative. But qualitative monitoring carried out in TEK combined with past experience allows accurate forecasting of ecological change and provides the opportunity to change behavior to fit new conditions (Berkes *et al.* 2000).

In the past, Western scientists frequently dismissed TEK as being a load of superstitious old tosh. Now there is sometimes a tendency to accept any claim based on TEK as valid and superior to the findings of Western science. Both attitudes represent failures to fully understand and value what TEK has to offer. Practices based on TEK must be tested just as any method deriving from scientific study must be tested (Woodworth 2010). In both cases, testing sometimes reveals that the proposed method was correct, and sometimes we find that it did not work. Because both TEK and science are based on observation of nature and empirical tests, both benefit from continued testing and refinement. If we treat TEK with the same respect and subject it to the same rigorous critique as we do Western science, then we will begin to fully appreciate and understand what TEK has to offer.

There have been occasions when I felt that TEK was limited to indigenous people, but Berkes and colleagues open TEK up to include anyone with long-term residence, close observation of nature and the acquisition of ecological knowledge. Thus they describe TEK practiced by caboclo and caicara people in rural Brazil, who have mixed racial heritage and practice ecological management based on both native and European traditions. They also report TEK among Icelandic fishers and Maine lobster and soft-shell clam fishers, because in those cases local people follow traditional custom to limit the number of fishers, the catch and thus preserve the fishery for all (Berkes *et al.* 2000).

I often wonder, how long does it take to become local, native to a place? When I was in graduate school in Rhode Island, living on Narragansett Bay, it was clear that I would never be considered native to Narragansett because my accent was wrong and marked me as a newcomer. But once I had spent a winter there and it was obvious I wasn't just a tourist, then I was accepted as belonging. I wasn't native, but I could fit in. Perhaps if I had stayed and had a family there, my children would have become native. As it is, I returned to western Illinois, where I am at least the fifth generation of my family to live, my children are now the sixth and where I feel as native as I will ever feel anywhere. Do I possess TEK? Perhaps a little about some things: I know when and where to find morel mushrooms, a skill passed down through the generations; I can feel the weather changing and know what the change means. But in many ways my TEK is rudimentary and impoverished.

How long it takes to become native and how long it takes to develop TEK are not trivial questions. The same questions apply to non-native species found in our rapidly developing novel ecosystems. At the current time we consider any species introduced by humans to a new area to be non-native. Thus cattle egrets are called native because they flew to the Americas on their own in the late nineteenth century. But nightcrawler earthworms (*Lumbricus terrestris*) are non-native because they arrived in the Americas via human transport in the seventeenth and eighteenth centuries. Yet nightcrawlers have so thoroughly altered North American soil ecology that they are now completely integrated into that ecology. Surely they must be native by now? And what about us? People are remarkably mobile thanks to modern technology. We move constantly, especially from rural to urban areas. We move from continent to continent. Once we have moved can we ever be native again? Can TEK be passed on from long-term residents to newcomers? If TEK is to survive and be incorporated into renewed restoration, we have to hope that modern people can become native to places and that TEK is transferable from resident to newcomer.

What kind of science best fits renewed restoration?

Tony Bradshaw famously (at least among restoration scientists) claimed that restoration is the acid test for the science of ecology (Bradshaw 1987). Bradshaw made that claim for two reasons. The first was to challenge ecological scientists to get involved with the highly practical and often difficult work of ecological restoration. Bradshaw published his remarks when I was in graduate school at the University of California, Berkeley. At that time, among the academic ecologists at that place, there was certainly a hierarchy of perceived worthiness of ecological disciplines. The pure study of ecology in the abstract or in ecosystems undisturbed by humans was considered superior to the applied branches of the field such as forestry, range management, fisheries and wildlife management, which were denigrated as simply being about ways to increase production of various commodities, whether it be board feet of lumber or salmon in a stream. Bradshaw was swimming against that tide and encouraged his colleagues to leave the purely abstract and get engaged with the profoundly important work of repairing our damaged ecosystems. The second reason for his remark was because at that time, ecologists had spent almost the entire twentieth century developing more and more sophisticated abstract theories and models about how ecosystems function, but had spent precious little time actually testing whether those theoretical functions were in fact important in real ecosystems. Bradshaw argued that if we could take a broken, damaged ecosystem and rebuild it based on our theoretical models of ecosystem function, then we would have real evidence that we truly understand how ecosystems work. Bradshaw was successful in his appeal to get ecologists to test their theories during restoration, and succeeding generations of restoration ecologists have done just that.

But have the restorations conducted in the past 25 years demonstrated that our scientific theories do apply to the real world? Has the practice of ecological restoration been improved by the rise of restoration ecology? Many, if not most, restoration ecologists would say yes it has – but then that is not surprising given that their *raison d'être* is using ecological science to improve restoration practice. Others have been less sanguine. My recent survey of the restoration community found that many practitioners are skeptical of restoration ecologists – especially restoration ecologists located in academia – because they think the scientists spend too much time testing abstract theory and fail to understand how the real world works. Robert Cabin expressed frustration with his own experiments in restoration and claimed he achieved more success when simply going out and restoring ecosystems. He felt that carefully designed experiments were too small and time-consuming to produce much restoration work of value and that often his experiments were testing ideas already known by practitioners to work (or not) (Cabin 2007, 2011).

In a somewhat provocative essay, Vidra and Shear (2010) asked whether ecological restoration is an art, a practice or a science. While they identify elements of all three in ecological restoration, they felt the science was often lacking in both applicability and success. Echoing Eric Katz's critique of restoration in general, Vidra and Shear asked whether restoration as the acid test for ecological science was really just "the next big lie." They pointed out that ecosystems are so highly variable due to climate, soil type, physical processes, species composition and past history that methods that work in one system may not work in another. Thus tests of general theory are bound to produce equivocal results because field tests will never be as clean as laboratory experiments – a situation that Bradshaw acknowledged was problematic in his original proposal of restoration as the acid test (Bradshaw 1987). Vidra and Shear also criticize restoration ecologists for only testing a small set of potentially applicable ecological theory and thus finding few theoretical ideas that are generally applicable across restorations. Assembly rules are one of the few theoretical ideas to be extensively tested and found to be important in virtually all restored ecosystems (Temperton *et al.* 2004).

In many ways the questions about what value science brings to restoration come from a mistaken understanding of how science works and the differences between pure and applied science. As noted earlier, when I was in graduate school I was trained (as were most ecologists of my generation) to think that pure science – conducted simply for the sake of gaining more knowledge due to our curiosity about how the world works – was superior to applied science driven by the need to produce a solution to a particular problem vexing humans. Despite that training, for many years I have been thinking the division between pure and applied science is a false dichotomy at best and in general it is not helpful when it comes to either developing theories or solving problems. But I was never quite certain how to express my misgivings until one afternoon in November 2010 I happened to be listening to BBC Radio 4 and heard an exchange that shed new light on the subject. A program entitled *The Infinite Monkey Cage* was on the air. The program is hosted by noted physicist and populizer of science, Brian Cox. In this episode, Cox was having an animated discussion with a guest about the value of pure versus applied science. In the end Cox exclaimed "The way you describe pure science sounds like applied science without the usefulness!" Cox went on to say we developed the laws of thermodynamics because we discovered them during our attempts to build better engines. We don't have better engines because first we had a theory of thermodynamics and then applied it to building engines. Instead, the discovery of thermodynamics was a by-product of our efforts to build better and better engines. As I reflected on restoration ecology I realized it was Bradshaw's background of restoring open-cast mines and old china clay pits that allowed him to see patterns that led to his general ideas about restoration as a way of speeding up the natural succession process.

That was when I realized why restoration scientists and practitioners sometimes fail to understand each other. Developing theories in the abstract is well and good, but we shouldn't be surprised that when we test those theories in the real world, most of them fail to describe what actually happens. Biologists are sometimes described as suffering from physics envy. We have witnessed the birth of quantum mechanics and seen how a purely abstract field came to both unify the discipline of physics and to have practical significance. We long for a similar unifying theory in biology. But what if the incredible developments in physics at the beginning of the twentieth century were a one-off? What if that is not the way science usually develops? Instead we work and work to solve particular problems – perhaps restoring ecosystems – and through improvements in our abilities to restore ecosystems we develop a new theoretical understanding of ecology. Rather than theory coming from abstract ecology to help solve restoration problems, instead knowledge gained while solving problems will flow into abstract ecology and benefit the growth of the science. Assembly rules have always been about solving particular problems – how do species colonize new habitat, how do ecosystems reform after a major disturbance like the eruption of Krakatoa – and from there led to more abstract theory.

Restoration ecology is an inherently challenging, observational, empirical science. Each ecosystem is unique. Solutions that apply in one may not apply in another. Trial and error, experimentation and repeated tests have allowed us to develop guidelines that work in particular ecosystems. It is also a historical science – we build on past successes and failures as we piece together the best practices for ecological restoration. We are starting to discern underlying principles that operate in all restorations. Restoration ecology and TEK both share a basis in empirical observation and repeated tests and in both historical memory – whether passed along in scientific journals or through oral tradition – allow refinement of methods and the development of a general understanding of how the world works.

Art in renewed restoration: helping bridge the gap between humans and the environment

When people ask whether ecological restoration is an art, they are referring to art in the sense of a skill that may be improved through practice and the application of certain principles or techniques, rather like the art of hitting a baseball or fly-fishing. They are not asking whether ecological restoration is an art in the sense of an activity that rests on creative imagination and intuition based on a sense of esthetics (Halle 2007). (Although I often argue that science is dependent upon creative imagination and therefore Einstein and Picasso have more in common than not, we generally understand art to be primarily the realm of imagination and intuition whereas science is grounded in empirical observation and experimentation.) But can or should art in the sense of the use of creative imagination, intuition and an esthetic sense play a role in ecological restoration? Art has tremendous, underutilized potential to help bridge the gap between human experience and the environment. Art can also play a role in allowing people to explore the meaning of ecological restoration and how they relate to both the process and product of restoration.

Art and the conservation movement have a long history of mutual interaction. An emotional affinity for nature is an excellent predictor of whether someone will work to conserve or restore wild nature. That emotional affinity can come from either direct experience with the environment or via art – whether the visual, literary or musical arts. There are many cases of art, such as the landscape paintings of Thomas Moran or the photographs of Ansel Adams, directly inspiring people to work for the preservation of nature, even nature they have never experienced in person (Curtis 2009).

Just as ecological restoration has been growing and developing over the past 30 years, so has a relatively new area of art – ecological art. The distinguishing characteristic of ecological art is that living ecosystems are part of the art. Sometimes ecological art has an inherent component of ecological restoration. Ecological art may respond to the environment, celebrate the environment and/or be set in the environment. Ecological art is often about expressing connections between humans and the environment, but at times it may also serve an instructional purpose and inform us of our role in the environment or how the environment develops over time (Curtis 2009). Some examples will help illustrate the range of art projects associated with ecological restorations.

The Illinois River in central Illinois was once home to extensive wetlands and supported huge populations of migrating ducks, geese and shorebirds. During the early to middle twentieth century, most of those wetlands were diked and drained in order to support large-scale agriculture. This huge wetland ecosystem was fragmented, destroyed and the populations of migrating birds were decimated. Loss of the wetlands also led to massive problems with flood control, erosion and declines in water quality. Now there are attempts to restore large areas of agricultural land back to wetland. One of the largest projects is at the Nature Conservancy's Emiquon Preserve. The goal at Emiquon is to restore about 3,200 ha of wetland. The project began in 2000 and from the beginning the Nature Conservancy's site managers wanted to include the public in the project as much as possible. To that end, a group called the Emiquon Corps of Discovery was formed in 2004. Their goal was to recruit members of the public (45 in the first group) who would observe the restoration process and respond to it artistically, whether by drawing, painting, photography, writing or however best suited the artist. The Nature Conservancy established a network of Aesthetic Points and Pathways that artists would revisit so they could use their art to both record and respond to the progress of restoration (Jeffords and Post 2005). The Emiquon Corps of Discovery were set up to function in an information-gathering, instructive way so that others not involved could observe the restoration of the wetland. The artists directly involved will almost certainly form connections to the place and hopefully their work will convey the sense of connection and meaning to others not directly involved in the project.

This kind of recording, instructional aspect is probably the most common relationship between art and ecological restoration. In Argentina the Ala Plastica artistic group used photographic documentation and mapping to record the scale of the 1999 Magdalena oil spill along the Rio de la Plata. The impacted area was part of a UNESCO Biosphere Preserve. Ala Plastica's presentation of the devastation encouraged members of the public to demand restoration of the site. Ala Plastica continued the project during restoration so that they could document the recovery progress and reflect on what was lost in the oil spill, what was recovered via restoration and what the newly restored ecosystem returned to the reserve. Similarly, Richard Kirk Mills has used print-making, digital archives of images and maps to show the effects of unplanned growth on remaining wild and semi-wild land in the greater New York metropolitan area. For six years he was the artist in residence at the Teaneck Creek Conservancy in New Jersey, documenting the restoration of a former landfill site. He participated in the restoration project so that he could learn about restoration and thus better educate the public via his art (Ball 2008).

Some artists have begun to use art as an integral agent of ecological restoration. Jackie Brookner makes living biosculptures from earth, pumice, bacteria, mosses and higher plants. Her biosculptures purify the air and water that wash over them. Her goal is to make larger and larger biosculptures that can function as integral parts of the ecosystem. She wonders why in many cities we have large drainage systems that directly flush stormwater away from the city, in effect wasting a vital resource. She would like to see sculptures and installations set up to capture and purify the stormwater, thus revealing the hydrology that underlies our built environment and provide an artistic expression of a necessary ecosystem function (Ball 2008). Peter Latz's design at Landscape Park Duisburg-Nord combines the artistic and scientific to create a park that provides ecosystem function and services, chronicles the history of land use at the site, creates esthetically interesting sightscapes and invites visitors to engage their sense of wonder as they explore the park.

Art provides a unique and accessible window to ecological restoration. The science of restoration is seen as objective and a bit distant because of that objectivity. Some people will be excited by the science, but mostly they are excited by the product of the science, the restoration in progress or completed. Because art and the response to art are understood to be subjective, the presence of art in ecological restoration invites human engagement. My experience is that the response is not always positive – in fact it is more often curiosity and puzzlement. Sometimes the response is negative. But in all cases it causes people to see and think about the restoration in a new way. And for me, as a site director, getting people to really look hard at and think about restoration is vital to the success of the restoration.

Case study: Blind – an installation at Green Oaks

For several years I have been working at Knox College's Green Oaks Field Station with environmental artist and Knox College art professor Tony Gant. I say working with him, but in reality my role has been mostly to provide him with space and permission to work, and then get out of his way. In 2006 he used an abandoned cabin to create an installation reflecting the history of human use of Green Oaks. When I first walked into the now transformed cabin, I felt like I was stepping into a gigantic version of a Joseph Cornell box. It captured the essence of people at Green Oaks in a playful but also cautionary way. It was powerful and added a new dimension to my understanding of the place.

Gant's more explicitly environmental work at Green Oaks began in the spring of 2008 when he was teaching a class based there and realized the site would be a good location for an installation set directly in nature. He observed me engaged in my on-going attempts to eradicate black locust from the restored prairies at Green Oaks (Allison 2011). He noticed that I was cutting down huge numbers of black locust saplings and wondered if he could use the downed saplings in his work. I said he was welcome to all of them he wanted; thinking the piles I had created would be more than enough. He then asked if I minded him cutting down more black locust as part of his installation and I thought "This is great – I can get someone else to help me deal with this invasive species." But I also wondered what kind of installation he had in mind. How large did he think the installation would become? He quickly sketched out a diagram on a piece of notebook paper and indicated the installation would go in a corner on the edge of West Prairie. It all looked fine to me, so I gave him the okay and let him go to work.

Throughout the spring and summer Gant was busy cutting down black locust, bundling the saplings into faggots and placing the faggots in various piles and structures. He also painted the cut sapling stumps in fluorescent orange and pink paint, often leaving stumps 1 m tall. His installation site expanded and grew beyond a corner of the prairie. Eventually he set up a huge simulacrum of a split rail fence, created from faggots of black locust saplings. He named the installation Blind and has continued to work on it ever since – sometimes adding new structures – now mostly deeper in the adjacent woods – sometimes taking things away (Figure 7.3). In 2009 I was scheduled to burn West Prairie, where his installation lay. We both thought that when the prairie fire struck the installation it would result in a huge conflagration – a prospect Gant considered exciting and which worried me. What we both failed to take into consideration is that black locust is extremely dense wood and a prairie fire moves fast – so fast it sped right over the installation and very little of the dense black locust actually burned.

I have been intrigued by how students, faculty, staff and visitors react to Blind. Many seem bemused by the structure. Most are curious about it, find it a bit puzzling and wonder what it is supposed to mean (which I will get to in a moment). Some comment that the installation is not esthetically pleasing. What is esthetically pleasing is extremely subjective – what pleases one

Figure 7.3 'Blind' is an artistic installation at the Green Oaks Field Research Centre. It is an on-going, evolving installation constructed by Knox professor and environmental artist Tony Gant.

Notes

This photograph was taken in early April 2010. There is restored prairie in the foreground, with faggots of black locust stems lying on the prairie ground. In the background is a row of black locust stems which have been cut off about 1 m above ground, treated with herbicide and painted with stripes of white and fluorescent pink paint. The black locust trees are mainly growing along the woody perimeter of the prairie and must be removed as they encroach on the restoration.

person may not please another. I suspect that Gant is not concerned with whether his work is pleasing or not – it should fit some esthetic principles, but pleasing is not part of the equation. A few – at least only a few have approached me – have been outraged by the installation. To them it is not art at all. It is a structure that disturbs their view of the prairie. The colors on the painted stumps are garish. It is not something that belongs in nature.

What does it mean? Gant is not an artist who is given to explicitly stating the meaning of one of his installations. He prefers to have the viewer ask questions, think about what is presented and come to their own conclusions. I'm sure it has a meaning to Gant and if you ask enough questions and offer enough interpretations, he slowly reveals what he was thinking by making somewhat enigmatic koan-like statements, but he won't come right out and say what it means. Art should make the viewer work and he isn't about to do the work for us.

This much I know: Blind asks us to look deeply and carefully at what happens to an ecosystem as it is restored. Restoration is not a simple process of positive, benign actions creating a new ecosystem. Like many creative processes, restoration has destruction embedded in it. We have to remove a lot of one unwanted species to allow the growth of the species we do want. That destruction has consequences and we need to carefully consider the consequences of our work. In his book *The Sunflower Forest* William Jordan describes a somewhat similar installation at the University of Wisconsin Arboretum Curtis Prairie. In that installation Barbara Westfall used the removal of invasive aspen trees from Curtis Prairie to ask questions about restoration. The aspen were girdled, revealing a rust-orange inner bark; Westfall then painted the girdled trunks with a mixture of vegetable oil to heighten the color and added black paint to outline the wounds. For Jordan (2003), Westfall's installation is about a ritual sacrifice and expresses shame and guilt about the necessity of the sacrifice. I don't think guilt enters into Gant's meaning; rather he wants us to see what is happening and also how from the destruction something new can be made. From the ashes, the Phoenix rises again – from the end of the black locust, the prairie returns.

Renewing the human community

If ecological restoration is as much about restoring the human connection to nature as it is about restoring ecosystems, then how do we get more people to engage with restoration? How do we generate more public interest and approval of ecological restoration? Often it seems that if we can simply get people working on a restoration project they see its value and are convinced that restoration is a worthwhile activity. If things go well, they then go out and encourage their friends and acquaintances to become involved and the circle of restorationists expands like the ripples on the surface of a pond after a pebble has been dropped in. But is that kind of informal expansion of the restoration community enough? Will the resulting community be robust and active enough to make a large difference in the quality of our ecosystems and the relationship of humans to them? Informal expansion of the restoration community can be effective at creating strong bonds between humans and the environment, but it is a slow process. Creating a truly robust restoration community will require more active nurturing and guidance. We cannot simply sit and wait for people to come to us with a desire to get involved – we have to seek out and invite more people to join our community (Scott and Whitbread-Abrutat 2010).

Several restorationists have commented on the need to develop ceremonies, traditions and rituals centered on restoration to both attract more restorationists and to reinforce the community already developing among restorationists (Cline 1994; Holland 1994; Meekison and Higgs 1998; Higgs 2003; Jordan 2003). Traditions develop within a particular community and thus are context-specific and meaningful to that community. Sometimes traditions develop gradually over time through repetition of a regular set of events, and the mere act of repetition provides meaning and focus for the participants. Some of the best examples of developing restoration traditions have occurred in the large, active restoration community found in the Chicago area. The North Branch Prairie Project has a traditional start to each working day that is so well established it has become a ritual that initiates a mental transition among the participants so that they are no longer preoccupied with their everyday lives, but instead become immersed in the day's restoration work. The ritual is simple and primarily concerned with the job at hand, but it is effective and serves to bring the community together. The work group gathers in a circle in which they go through a greeting process – each introducing him/herself, relating any concerns, exchanging information about the progress of the restoration and planning the day's work (Holland 1994). There is nothing complicated about this ritual start to the day's work, but it creates a space in which people can shift gears from the daily humdrum and instead focus on a different set of concerns.

Ideally, restoration rituals would all develop organically from local traditions like the North Branch Prairie Project did. Ritual, in restoration and elsewhere in our societies, performs several functions. On one hand rituals create conditions in which individuals and communities seek and accomplish change in their values, perceptions and even their way of living. It does so by allowing the discovery of new relationships and the re-examination of existing ones. On the other hand ritual may also act to reinforce social norms. Sometimes it can do both – ritual initiates new members to group practices and attitudes, thus exposing them to new ideas and facilitating change and at the same time it renews and reinvigorates established members (Cline 1994; Meekison and Higgs 1998). Thus at North Branch the gathering ritual at the start of the day introduces new members to the workings of the restoration group and reaffirms the process for old-timers.

In our increasingly fast-paced and fragmented societies, ritual is sometimes viewed with suspicion. Some feel ritual harkens to a more superstitious, old-fashioned way of life or that ritual is exclusively the province of religion or organizations like the Masons. There are restorationists who have spoken of ritual in restoration in quasi-religious terms. William Jordan in particular would like to see ritual restoration form a new communion with nature. In his view this communion would both renew our relationship with nature and also allow us to atone for the shame and guilt we feel at the damage we have caused to the earth (Jordan 2003). Jordan specifically used the word communion because of its religious overtones (especially for Christian communities) and its notion of communing – an intimate sharing of thoughts and feelings. While such a communion is an admirable idea, I have observed that it makes many fellow restorationists uncomfortable. Some have said that they already have religious beliefs that work well for them and restoration is not likely to supplant that. Others coming at restoration from a strictly scientific background have little interest in religion of any kind. In either case, imposing a quasi-religious ritual on restorationists runs the risk of turning people off to restoration and reducing the community of restorationists (Meekison and Higgs 1998).

Eric Higgs prefers to think of the developing restoration rituals as focal practices – activities that "encourage the development of the skills necessary for competence and excellence" (Higgs 2003, p. 190). When performed well, focal practices allow a group to come together to pursue an agreed-upon goal in an enjoyable manner. While I appreciate Higgs attempting to develop a more neutral term for ways to reify restoration practices, focal practices sounds too clinical to me. I prefer to think of them as traditions, but whatever term is applied, the key is that these traditions initiate new members in the practices and norms of the group and reinforce the values shared by group members. Ann Cline thinks that one of the most positive aspects of rituals/traditions is that they force the participants to slow down. Slowing down does two things – first, it helps pull us away from the pressures of modern life; and second it gives us time to enjoy the moment, even to pause and enjoy the moment of the mundane. She uses the example of the Japanese Tea Ceremony as a ritual that slows down the normal, simple task of making a pot of tea so that in the extended time of the tea ceremony each step can be savored and reflected upon. For her, restoration rituals should not discuss the ecological function of the restoration, but rather should focus on the pleasure of the activity (Cline 1994). Focusing on the pleasure of restoration is somewhat at odds with our usual desire to explain the ecological value of our work, but my own experience tells me that it is the pleasure of some kinds of restoration that form the initial basis of the restoration community.

Every spring, usually in early April, at Green Oaks we burn one of our restored prairies. As the director of the field station, I am in charge of maintaining, and to a certain extent reifying, the traditions that have grown up around the annual fire. The annual fire is a big event and a major part of the identity of Knox College, so major that the school's athletic teams are named the Prairie Fire. Traditions about the prairie fire were in place before I was hired and for the most part I have continued them. A week or two before the fire I send out an announcement of when the fire will be held, weather permitting. On the day of the fire interested members of the Knox community – primarily students – come to Green Oaks to assist with fire control. I hand out fire-fighting equipment – flappers, fire mops (burlap attached to long wooden handles, the burlap soaked in water so it can be used to snuff out small fires) and backpack water sprayers. Then I give a fire safety talk in which I emphasize the dangers of fire and the necessity of treating the fire with the utmost respect and wariness. I often read a quote about prairie fires (the quote used varies from year to year and has come from George Catlin, Aldo Leopold and John Madson) to lend an air of grandeur to the event. Then we go and burn the prairie. Once the prairie has been burned and all fire extinguished, we return to Schurr Hall at the field station for a picnic –

traditionally hamburgers and soft drinks. Many of the students walk across the burned prairie after the fire and daub themselves with ash in homage to the fire and an unconscious invocation of Ash Wednesday.

As far as I can tell, my predecessors at Green Oaks were comfortable with this tradition. But the more times I conduct prairie burns, the more nervous I get about them, the more I see the potential for something to go badly wrong. When I first arrived, I would have about 30 students show up to help with the fire. However, in recent years the number of interested participants has grown to over 100, which cause a number of problems.

I don't have enough equipment for everyone to participate in fire control; I don't have a budget to feed that many people; and I have a hard time managing the fire and practicing crowd control at the same time. So I have attempted to change the tradition to reduce the number of participants and restrict other people to the role of observers from a safe distance. I have also had to reduce the picnic to soft drinks and snacks. These changes have not been popular and I don't like them either – it changes the entire feeling of the tradition. If restoration is to renew the human relationship to nature, it should not be exclusive to a few hand-picked individuals. The meager provisions lessen the sense of celebration once the fire is over. And yet such changes are necessary for the safety of everyone at the prairie burn. Fire is a sexy beast – exciting, seductive and very, very dangerous. We cannot be too careful when we use it. Smaller numbers are safer. From the perspective of generating excitement about restoration, nothing works much better than letting someone tend a fire line. The thrill of watching the prairie burn and standing there in the smoke and blowing ash as the fire races across the prairie is exhilarating. Not everyone who comes to watch or help with the prairie fire will return to do more restoration work; some are just there for a grand afternoon's entertainment. But enough see what restoration is really about and keep coming back to do more, less exciting work. They become committed to restoration and often continue working with restoration – either as professionals or volunteers – long after graduation. In contrast I have only limited success at turning initiates into excited, dedicated restorationists when their first experience is coming to Green Oaks to cut down brush or pull weeds. Such work is necessary, but it often feels like drudgery and it doesn't have the same attraction as working with fire. With my students, a bit of excitement and pleasure at a job done well is what turns a new volunteer into a committed restorationist.

Another approach is to create a ceremony to celebrate the hard work done earlier in the year. One of the most successful such events that I know of is the Bagpipes and Bonfires Celebration held in the autumn in Lake Forest, Illinois (www.lfola.org/events/bagpipesandbonfire.aspx). Lake Forest has a savanna restoration site in the Chicago area. Restoring savanna almost always requires us to cut down much brush when removing invasive species. The detritus and remains from all that removal is piled up and saved for a celebratory burn in the autumn. The bonfire is accompanied by a picnic, children's games and music from bagpipers and is usually attended by over 1,000 people. It combines a celebration of savanna restoration with a celebration of the heritage of the Scots-American founders of the town (Holland 1994). Celebratory traditions that have arisen in a particular place result in meaningful ceremonies that rejoice in the special relationship between people, nature and restoration. Those ceremonies are sustaining and inviting and likely to foster the growth of the restoration community precisely because they are fun and not threatening to people inexperienced with restoration. Once they have come to the party, seen that the restorationists are their neighbors and are just like them, it becomes much easier for them to see themselves also helping with restoration. And once they have the satisfaction of actually working on a restoration, of getting out and sweating a bit and seeing the difference that a few hours of work can make, then they may well become active members of the restoration community.

Conclusions

Today we have an urgent need to get out and restore damaged ecosystems. A global restoration program is of paramount importance both to stem staggering losses of biodiversity and ecosystem functions, and to renew the bond between humans and their environment. Yet despite our sense of urgency, we must be humble in our approach to restoration. We have to work with what we have – the damaged ecosystems scattered across the earth, the species (native and not) found in those ecosystems, the coming effects of global environmental change and our limited knowledge of how to best restore those ecosystems. Restoration has to be done at a global scale in order to achieve its potential to be the most beneficial way we can respond to our damaged earth and the increasing pace of environmental change. But we know it is easiest to perform restoration at the small, local scale (Ling *et al.* 2003). Somehow we must start at the small scale but within the context of a much larger vision. Our best model of how to proceed is the restoration and preservation of the European Green Belt. And that is cause for optimism. If the nations involved in the EGB can find ways to overcome old enmities and distrust to work together, to reach out to individuals and small communities while working within a multinational planning framework, then there is hope that we can do the same across the globe. A renewed restoration program can restore ecosystems and human communities, but the greatest benefits of such a program will only be realized when it combines individual ingenuity and communal effort with a global perspective. We must hope that we can create that program before the twenty-first century gets much older.

8 Building the restored future

Making the renewal happen

In this chapter I will:

- return to the theme of embracing the role choice in restoration that I first explored in Chapter 1 and will quickly summarize some of the challenges facing restorationists in the twenty-first century;
- respond to some criticisms of ecological restoration;
- examine ecological restoration as a way to achieve ecological integrity, historical fidelity and complete the search for authenticity in managed ecosystems;
- discuss the need for a new, practical ethics of ecological restoration;
- acknowledge the limits of what ecological restoration can achieve;
- offer six practical steps all restorationists should take;
- end with a note of optimism.

Back to choices

I started this book with a discussion of the role of choice in ecological restoration, and in many ways my comments about choice in restoration apply to human choices in any form of environmental management. Everything we do in the environment comes about due to choice. Even choosing to do nothing in the environment (which in some cases such as managing larger wilderness areas may be the best choice) is a choice. Given the size of the human ecological footprint, it is inevitable that anything we do or do not do has consequences for the environment. Thus there is no way to avoid making choices – as my professor Frank Cousens told me so many years ago, failure to make a choice is a choice. Not making a choice is usually a choice to continue with the status quo. And the status quo in terms of human use/abuse of the environment is not a choice we can afford to make. Our ecological footprint continues to expand, and with it we see increasingly negative trends due to a continuing loss of biodiversity, fragmentation of ecosystems and output of environmental pollutants. We need to acknowledge the global size of our ecological footprint and embrace the serious consequences and responsibilities that have fallen to us in our role as environmental decision-makers if we are to have any chance of changing our behavior from the status quo to a more sustainable way of life.

Talking about the reality of our role as environmental decision-makers is controversial both to those who prefer to think of nature (or some other entity bigger than us) as the real decision-maker (and therefore think we have no right to make decisions about/for the environment) and to those who think the status quo is just fine and see no need to change our behavior. We no longer live in a time when we can sit back being demure about our role as decision-makers just to avoid controversy. We have to engage the controversy and discuss our role as decision-makers even if the main outcome of that discussion is simply to get people – all people – to think more seriously about our position in the global ecosystem and our responsibilities to ourselves and the entire environment.

Although my literature survey revealed geographical differences in people's goals for restoration projects, my online survey found an emerging consensus that ecological restoration is vital for maintaining (and hopefully increasing) local biodiversity and ecosystem function. Restorationists around the world may differ in specific goals for their work, but in general they increasingly agree about core values of restoring biodiversity and ecosystem function. To me, this developing agreement is cause for optimism because if we are in broad agreement we should be better able to make consistent, sensible choices about how and when to engage in ecological restoration.

In previous chapters I discussed the evidence that we are now living in a time of increasingly rapid and large environmental changes. Because these changes are frequently driven by human activities, some scientists have suggested that we are now living in a new geological epoch which they have named the Anthropocene in order to recognize the amount of influence we now exert at a global level (Crutzen and Stoermer 2000). It is clear that during recent human history we developed technologies that allowed us to both increasingly contour the planet to fit our needs and desires and also to free us from many of the whims of nature (although recent earthquakes, tsunamis, volcanic eruptions and violent storms remind us that Mother Nature will always be a formidable force in our lives, able to wreak havoc on even our most highly engineered projects). We have reached a point at which our behavior exerts a considerable influence on global environmental processes and unfortunately that influence is frequently detrimental to both us and the rest of the earth.

Global environmental change is nothing new – we know that in the recent geological past the glacial epochs were periods of constant environmental change. The past 20,000 years have seen continuous, directional environmental change resulting in a steady increase in global average temperature (Jackson and Hobbs 2009). But the pace of change is rapidly accelerating. During the last 100 years the earth experienced more rapid increase in temperature than during the previous 100 years. Models predict that the next 100 years – our own twenty-first century – will experience even quicker and greater magnitudes of change than ever recorded (Lawler *et al.* 2010; Dudley 2011).

Thus far I have focused on environmental change, but we also need to consider – if only briefly – equally swift changes in the way we humans live. Since the start of the Industrial Revolution, and especially in the last 100 years, there has been a drastic change in how and where humans live. Until the Industrial Revolution most humans lived in rural areas which may have been domesticated, semi-wild or wild, but from which most humans derived all the products needed for their sustenance. After the Industrial Revolution there was a rapid shift of population from rural areas to urban areas, which are increasingly becoming vast megacities with huge populations. By 2008 over 50 percent of the human population lived in cities and that percentage is expected to continue to rise to more than 60 percent by 2030. In the developed world there has been an especially large shift to urban areas so that 80 percent of people in developed nations live in cities and their suburbs. In the United Kingdom 90 percent of all people live in urban areas (Miller 2005; Goddard *et al.* 2010).

This momentous shift to urban living has resulted in what Robert Pyle has referred to as "the extinction of experience" (Miller 2005). More and more people live their entire lives in highly domesticated, simplified and managed urban areas. They have almost no connection to wild nature, biodiversity and open spaces. Thus the connections to natural ecosystems that have defined human experience for most of our existence are lost to them. E.O. Wilson has written about humans possessing biophilia, an evolved, innate love of biodiversity (Wilson 1984); but biophilia, like any kind of love, cannot flourish if the ground upon which love could grow is barren. If people's experience of biodiversity is limited to a few species that tolerate urban conditions – crabgrass, blue grass, dandelions, ailanthus, starlings, rock pigeons, crickets and feral cats – they are unlikely to place much value on biodiversity or understand the need for high levels of biodiversity. Similarly, if their experience of ecosystems is limited to vacant lots and channelized, concrete-lined streams, they will have a hard time understanding the complexity of ecosystems and the need to preserve ecosystem function. This extinction of experience may result in people either feeling omniscient with respect to environmental choices – a feeling that we can build whatever we want, whenever and wherever we want – or perhaps even worse a fatalistic acceptance that our choices won't matter at all, that everything will be the same no matter what we do. In either case, engaging an increasingly urbanized population with little experience of anything approaching wild nature will be one of the greatest challenges to achieving renewed restoration in the twenty-first century.

Responding to critics

If ecological restoration is to fulfill its role in the twenty-first century, restorationists must be ready and able to respond to our critics. The most serious criticisms come from within the conservation community and take a couple of different forms. One form, discussed in Chapter 1, is best exemplified by the arguments made by Eric Katz in which he questions the legitimacy of the entire ecological restoration program. For Katz, restoration is "a big lie" because ultimately it is impossible to ever restore wildness to a landscape. The act of restoration, because it is the result of imposition of human choice and will, robs the restored landscape of the continuity of natural (for Katz only non-human) forces and results in a form of domination of nature in a misguided attempt to repair the landscape. Katz argues that efforts put into restoration are wasted and that instead we should concentrate all of our energies on the preservation of remaining wild land (Katz 1992). When I talk with people about ecological restoration, I have run across some who make similar arguments and ask whether restoration in fact diverts funds and effort away from more important preservation work.

The response to a denial of the legitimacy of restoration needs to follow several lines of argument because the criticism proceeds along several different lines of reasoning. First I will go out on a limb and say that every restorationist also thinks that it is vitally important that we preserve all remaining wild land. I would be shocked if any restorationist disagreed with my claim. I will not claim that all restorationists put preservation of wild land at the top of the list of priorities – I know that would be incorrect – but no one would leave it off of the list. Second, we can also respond that if our definition of wild nature is nature absolutely untouched by human activity, then the game is already over – there is no place that is not influenced in some way by human activity, even if it is only (as if this is a small only) the effects of global climate change.

Third – and this is a more important response from the perspective of restorationists – is to ask what is meant by "wild" and why "wild" is destroyed by human touch. Nigel Dudley notes that for some preservationists, wild is a quality rather like virginity – once it is lost it can never be regained (Dudley 2011). In that formulation of wild, wildness becomes a thing, an object that is subject to debasement and defilement. But in ecological terms, wildness is not a thing, it is a process. And as a process it can be restored. As we will explore in more detail in a moment, Eric Higgs argues that the best ecological restoration incorporates wild design in which we purposely plan for ecosystem dynamics to help shape the development of restored ecosystems (Higgs 2003; Higgs and Hobbs 2010). In fact, any time we work with living organisms, we have introduced a strong element of wildness. Organisms will live, grow and reproduce (or not) in response to environmental conditions in ways that we cannot possibly control. As soon as we scatter the seeds, plant the plants, introduce the animals, microbes and fungi, and then step back, the wild creeps in and our control grows less and less. As restorationists we hope to set up conditions that will favor the development of a particular ecosystem, but ultimately the exact form that ecosystem takes will be shaped by ecology as much as by our plans. Thus wildness is always present. This is something that classical and renaissance gardeners recognized when they included areas of gardens that were intentionally left unmanaged, both to introduce the wild to the garden and to admit our humility by recognizing there are limits to our efforts to control nature. These *boscos* were often small, but they were central to the meaning and spirit of Italian gardens for many centuries (Mitchell 2001).

Criticisms from fellow conservationists sometimes take the form of criticisms of the science underlying ecological restoration. Again, I will go out on a limb and say that all restorationists agree that the scientific knowledge necessary to restore ecosystems is still being developed and thus could be better. But the criticism is often not just of the quality of scientific knowledge, but also of how that scientific knowledge is applied when restoring ecosystems. Davis and Slobodkin (2004) provide an interesting case in point. They examined the 2004 *SER Primer on Ecological Restoration.* Among the main goals listed in the *SER Primer* are restoring ecosystem health and integrity. Davis and Slobodkin claim ecosystem "health" and "integrity" are not scientific terms because they cannot be easily defined or measured. Ecosystems are not subject to natural selection because they are not individual organisms, and because they have not evolved via natural selection they cannot possess health or integrity. They claim that because health and integrity do not exist for ecosystems, those terms are value judgments and not scientific properties. As I will describe in a moment, there are ways to define ecosystem integrity so that it can be measured via scientific methods, and Parks Canada in particular has been successfully doing so for many years (Winterhalder *et al.* 2004; Woodley 2010).

I want to focus on the negative tone Davis and Slobodkin use when discussing values in the context of science. A few years ago, some of my students wore T-shirts emblazoned with the slogan "You say I'm a feminist like that's a bad thing." And I feel that Davis and Slobodkin are also saying "value" like it is a bad thing. For them science has to be completely objective and not muddled by the imposition of human values. Scientific questions are all abstract, the data speak for themselves and conclusions are thus divorced from human desire to interpret them in a particular way. It is possible that if scientists were Vulcans like Mr. Spock and came from a planet where logic reigns supreme, then science might work that way. But because we are humans, science can never be value-free. We make value judgments when we decide to study certain subjects – I found ecology more interesting than chemistry or physics, a value judgment on my part.

Scientists make value judgments when they decide which questions to pursue. Funding agencies make value judgments when deciding which grant proposals should be funded and which should not. Editors make value judgments when deciding which manuscripts to publish and which to reject. In an ideal world those decisions should be based on strict interpretations of the data and its presentation, but ultimately decisions are based on a set of values that are strongly influenced by each scientist's background, education, personality and social milieu. Our scientific knowledge will help us decide which ecosystems to restore and help us test methods to best restore them. Science is vital in our original decision to restore, planning the restoration, the implementation and management of the project – but that science is informed by human values at every step in the process.

Davis and Slobodkin seem to be confusing the science of restoration ecology with the practice of ecological restoration in their critique of the *SER Primer*. They appear to want the entire practice of restoration to be as objective and removed from value judgments as possible. However, the most important aspect of ecological restoration as a practice is the integration of many different modes of thinking and types of knowledge in the process of restoration. Scientific knowledge and good data are vital for successful restoration, but no more vital than social policy, political will, economic decisions, cultural values and ethical considerations. In many ways, the scientific questions about a restoration are the easiest questions to answer (although they can be extremely thorny). It is the social, cultural and political questions that pose the greatest challenges to ecological restoration. And these questions lie beyond the realm of science. Reducing ecological restoration to the science of restoration ecology is highly limiting. We might be able to accomplish restorations with a high degree of technical rigor, but we would be lacking the values and wisdom that are not usually considered a part of science. As Eric Higgs has said, the practice of wisdom will be as important as the practice of science in ecological restoration in the twenty-first century (Higgs 2005).

Thus restoration ecologists have to be well versed in the language of values, and comfortable talking about values. We need to be able to describe our own values, why we began this work, what we think restoration will provide to society and to the environment. We must clearly identify when we are speaking about values and when we are presenting as objective an analysis of the data as possible, but to deny that values play a role and to deny our own values is an abdication of our duties as scientists to provide the best possible interpretation of the data to the public.

Most scientists were educated with the support of government-funded programs and conduct their research using government funding. Thus we have an obligation to the citizen-taxpayers who supplied those funds to provide the judgment we have acquired as a result of that education and research. We live at a time when scientific expertise is desperately needed and desired to help people and society make decisions about how to face our environmental challenges. But in order to speak effectively to non-scientists, we must be able to relate our knowledge and values to the public in clear, everyday language that expresses what we know and, just as importantly, what we don't know. If we cannot communicate what we have learned in a respectful and well-spoken manner, we will fail to convince anyone to engage in ecological restoration (or any other conservation work) (Olson 2009; Groffman *et al.* 2010; Meyer *et al.* 2010).

Ecological integrity, historical fidelity and the search for authenticity

The 2004 *SER Primer on Ecological Restoration* lists ecological integrity as one of the key attributes of a restored ecosystem. It also discusses ecosystem health, which is harder to define, but a healthy ecosystem will have integrity so we can claim that integrity is a (perhaps *the*) key component of ecosystem health. Some critics of the concept of ecosystem integrity claim it can't be measured or defined, but it is already in use by several management organizations and listed as a goal in many policy statements besides the *SER Primer.* Among these are the 1978 Great Lakes Water Quality Assessment, the 1998 Parks Canada National Parks Act, the 2004 Statement by the Council on Biological Diversity and the 2005 Millennium Ecosystem Assessment (Millennium Ecosystem Assessment 2005; Woodley 2010). The adoption of ecological integrity as a management goal marks a large shift in perceptions of what ecosystem managers should be doing. Traditionally, most national parks and similar wild areas were managed in order to maintain naturalness (also a rather difficult term to define). In many cases naturalness was seen as the absence of human intervention in the ecosystem and thus implied that there should be minimal management of wild areas. Ecological integrity instead springs from the recognition of two facts. First, it became obvious that due to factors like global climate change, the spread of invasive species and increasing human pressure on the environment, that minimal management was failing and formerly wild areas were suffering losses of biodiversity and declines in ecosystem function. Second, a better understanding of ecological history revealed that in virtually all landscapes (with the exception of Antarctica and high alpine regions) humans were, are and will continue to be major players in ecosystem dynamics. Ironically we learned that in our current environment, wild areas require management in order to maintain any semblance of the wild features we valued when they were first conserved (Woodley 2010).

When we talk about something having integrity, we mean that it is whole and it behaves properly. Thus for an ecosystem to have integrity, it must all have its components (ecosystem structure is maintained) and must perform its functions well. Based on that we can establish measurable parameters such as expected species and trophic composition, functions that will occur within the ecosystem, expected levels (maximums and minimums) for structures and functions, and thus determine benchmarks against which we can test the current state of the ecosystem. Parks Canada has been especially successful in applying the principle of ecological integrity to its management of Canadian national parks. In Banff National Park, one of the flagship Canadian parks, mixed human- and lightning-caused fire regimes, predator–prey relationships and a replication of native people's hunting practices have all been implemented based on the idea of ecological integrity. The result has been an increase in biodiversity and ecosystem function in the managed areas (Woodley 2010). Because active human involvement is a key component of managing for ecological integrity, the concept fits in well with ecological restoration programs.

For many restorationists, the major component of structure that is of interest is biodiversity – in particular the species composition of the restored ecosystem. We are often concerned with a loss of rare species when we think about the need to restore biodiversity. However, restoration is often planned and focused on local sites. We might do an excellent job of restoring local sites and preserving local biodiversity, but biodiversity loss is a global problem and a focus on local sites may miss global needs. This is particularly true when considering rare species. Because rare species are often hard to find and enumerate (and thus poorly known), we may miss them when planning and carrying out restorations.

Our current focus on ecosystem function means that many restorationists attempt to restore a full range of functional diversity – so that all types of ecosystem function are restored to the ecosystem. A focus on functionality may result in ecosystems that function well and carry out desired ecosystem services but which lack rare species. It may be possible to restore full function with a suite of widely distributed, common species. Shahid Naeem and his research group are very concerned about maintaining and restoring phylogenetic diversity as a way to ensure that restoration of biodiversity includes rare species (Flynn *et al.* 2011). They have found that when rare species are maintained, more functional traits and more biodiversity are preserved. Restoring ecosystems so that we maintain historical continuity with the original ecosystem helps ensure that we capture rare species and a full range of biodiversity – and both functional and phylogenetic diversity – in our restorations.

Another cornerstone of the *SER Primer* is the idea that ecosystems will be restored to the historical trajectory they were on before disturbance and that ecosystems will have some level of historical fidelity to their past condition. An "ecosystem with historical fidelity ought to be compositionally, structurally and functionally similar to past ecosystems found at that place" (Cole *et al.* 2010, p. 127). Although not always referred to as historical fidelity, the idea that conserved ecosystems today should be similar to past wild ecosystems is a common component of most conservation and preservation programs. There are many reasons to pursue historical fidelity in conservation and restoration work. Some of them are primarily cultural – an attachment to a past ecosystem that has a particular meaning to a group of people. There are also good scientific reasons to desire historical fidelity – in particular the fact that if a current ecosystem resembles the original it is likely to have all the parts necessary for proper function. However, the dynamic nature of all ecosystems and the current rapid pace of environmental change make it increasingly difficult, even unrealistic, to base restoration and conservation programs on the goal of maintaining historical fidelity. Historical fidelity has to be a part of restoration and conservation because without some relationship to past conditions we are not really conducting restoration or conserving the ecosystem that we originally valued. The issue is the degree of historical fidelity that is required in any particular restoration. As Cole *et al.* note, today the concept of historical fidelity is best applied in a "relative rather than an absolute sense" (Cole *et al.* 2010, p. 128).

When trying to decide how to apply a relative sense of historical fidelity we have to ask ourselves what is most important, the components of the historical object/system or the function of the object/system or a combination of components and function? Frequently in Western cultures we focus on the components of the historical object or system. In 2002 I took a class to visit Aldo Leopold's Shack, the setting for many of his essays in *A Sand County Almanac* (Figure 8.1). For many years the Leopold family and later the current owner, the Aldo Leopold Foundation, were secretive about the exact location of the Shack somewhere in the Sand Counties along the Wisconsin River. They were secretive partly to protect the privacy of the Leopold family, who still use the Shack for family reunions and partly to protect the integrity of the Shack. There had been unfortunate incidents where Leopold fans (fanatics, really) had visited the Shack and broken off pieces of wood to take home as a talisman from their visit to the home of their hero. I had never seen the Shack before and was excited that the Aldo Leopold Foundation had agreed to take my class and me on a tour of the Shack and the property where Aldo himself had begun his early experiments in ecological restoration. When we got to the Shack it was thrilling to see the actual building where Leopold and his family spent so many happy hours. I was surprised to see that it really was a shack – I always thought the title was a self-deprecating family name for a cabin or cottage. But no, it was truly a small shack with an earthen floor, constructed from a salvaged shed and driftwood boards. Seeing that it was so rustic, combined with the opportunity to sit in Aldo's favorite chair, to recline on the simple bunks and rest by the hearth, made it even more thrilling. Sitting in his chair I secretly hoped that some of Aldo's genius and generosity of spirit would rub off on me.

But later, as I reflected on Leopold's Shack, I was taken back to an undergraduate course in which we discussed the paradox of Theseus' ship.

Figure 8.1 The author (far left) and other members of the 2002 Green Oaks Term visit Aldo Leopold's Shack in the Sand Counties of Wisconsin.

The ancient Athenians had preserved the ship on which Theseus and the youth of Athens returned to Athens from Crete. The ship sat out in a plaza and the weather took its toll on the ship so that some timbers rotted and decayed. As the timbers were destroyed by the elements, the Athenians carefully crafted replacements until, after many years, perhaps even centuries, the ship still sat in the plaza, but none of the original wood remained. Over time it had been entirely reconstructed from new materials. This reconstruction led the Athenians to ask whether it was still the same ship on which Theseus had sailed home. Well, was it? Frequently in the West our answer would be no, it may look like Theseus' ship, but it is not the same ship because none of the original wood remained. So if over time, overly enthusiastic souvenir hunters had gradually carted off all the original wood from the Leopold's Shack but the Leopold Foundation had assiduously replaced each lost piece (perhaps even using scavenged driftwood as had the Leopold family when making their own repairs), would the Shack still be the same one that Aldo had slept in? We Westerners would probably say no, it wasn't the same. And if it wasn't, would that matter? Is the Shack and the surrounding land important because of the exact components of the building that exist there where Aldo walked and slept, or is it important for the process that Aldo engaged in when working there?

In other cultures, people respond to that same paradox rather differently. In Douglas Adams' book, *Last Chance to See* (1992), he described visiting the Golden Pavilion Temple in Kyoto. The temple was built of wood in 1397 and Adams was amazed it was in such good shape almost 600 years later and expressed his surprise that it had withstood the elements so well. His guide told him it had not survived the elements at all – it had burned down and been rebuilt many times, most recently in 1950. Adams thought about this and said: so it isn't the same building at all. His guide responded that no, it is the same building; it is historically and culturally important so they rebuilt it and it is exactly the same building. For the keepers of the Golden Pavilion Temple in Kyoto, the important aspect is not that the temple is still constructed from the original timbers present in 1397; what is important is that the temple has the same form and serves the same function as it has since 1397.

And so we return to the relative sense of historical fidelity in restoration and conservation. Given the rapid pace of environmental change and the development of novel, no-analogue ecosystems, we cannot expect our restored and conserved ecosystems to have exactly the same species composition as before. We cannot always use the same pieces to rebuild what once was. But we can realize historical fidelity by maintaining the same form and function as previously existed at the site. The ecological processes are the critical feature of a restored ecosystem. We would like them to have as many historically important species present as possible, but we have to be realistic about what we can achieve.

In the twenty-first century, renewed restoration should strive to produce what Nigel Dudley has referred to as an authentic ecosystem (2011). He defines authentic ecosystem thus: "An authentic ecosystem is described as: *a self-regulating ecosystem with the expected level of biodiversity and expected complexity of ecological interactions, given historic, geographic and climatic factors*" (Dudley 2011, p. 138, emphasis as in the original). He admits that given current conditions and human pressures, self-regulating will be a difficult standard to achieve and later suggests that perhaps an authentic ecosystem should be one that functions with the minimum amount of human manipulation possible. When reading Dudley's work it is obvious that he loves wild ecosystems that have a long history of continuous existence in a particular place, but he recognizes that such places are becoming increasingly rare. Thus I find it interesting that when he discusses authenticity, he does not mention the identity of particular species – what makes an ecosystem authentic is its form and function. If the ecosystem, rather like the Golden Pavilion Temple in Kyoto, has the same form and function as the previous ecosystem, then it passes the authenticity test. I am glad that Leopold's Shack still stands in the Sand Counties beside the Wisconsin River, and I'm very glad I got to sit in Aldo's chair. But the most important aspects of the Shack are the ideas that were born there and the processes that were the fruit of those ideas. Even if the original Shack was destroyed and rebuilt like the Golden Pavilion Temple, its meaning would remain, the processes would continue and it would be the same building as before.

The size of the human footprint and our ability to manipulate the environment has, for better or worse, forced us to fully recognize and embrace our role as stewards of the earth's biodiversity and ecosystem processes. The key practical and ethical issue in this century will be for us to recognize that as stewards we have to decide how authentic our ecosystems will be (Dudley 2011). It will behoove us to be as authentic as possible, using native, historically present species as much as possible, but above all ensuring that the ecosystem functions that have maintained life on this planet for millennia continue long after we are gone.

For me the critical aspect of authenticity is that we must recognize our limitations in controlling the processes and functions at work in ecosystems. Dudley is right to be wary of calling for a strict reliance on self-regulation as a key marker of authenticity. The ecosystems that I work with – restored tallgrass prairies and Midwestern oak savannas – will almost certainly not achieve self-regulation. Even the largest are too small to maintain a wild fire disturbance regime and it is likely that both ecosystem types originated in the most recent post-glacial period and evolved during a time of constant human manipulation (especially by use of fire). Thus wild self-regulation without human agency would be unnatural for them. Some degree of human manipulation will always be important in maintaining authenticity in those ecosystems. But at the same time that manipulation must not be so heavy-handed that it precludes the opportunity for wild processes.

Eric Higgs coined the seemingly contradictory term "wild design" to describe a management process that explicitly plans for and supports wild, unregulated processes in restored and conserved ecosystems (Higgs 2003; Higgs and Hobbs 2010). Our planning must set up conditions in which species are free to interact and gradually establish their own patterns of abundance and distribution. Because ecosystems are composed of living organisms, they will always exhibit some aspects of self-design which we cannot control (Seddon 2010). Even highly damaged and heavily manipulated sites such as the Landscape Park Duisborg-Nord can be planned to allow for wild processes to operate and thus produce the ecosystem best suited for the new environment (Latz 2001). The component species in each ecosystem will interact in ways that are not always predictable, especially given the quickening pace of environmental change. Some currently dominant species may decline in importance to be replaced by new dominants. Predator–prey relationships, competitive interactions, host–disease relationships and mutualisms will change over time as populations wax and wane. New species, some not terribly desirable, will disperse into restored and conserved ecosystems. But as long as the overall form and function remain authentic, the ecosystem will have integrity and the restoration will be successful. The truly difficult challenges for us as stewards will be when environmental change and wild processes cause an ecosystem to change form – perhaps grassland is transformed into savanna as the new climate favors trees or a permanent change to drier conditions results in a lowering of the water table and a loss of wetlands. In those cases we will have to ask, what can we achieve at this site? Is it possible to maintain the grassland or wetland? Or is it time to facilitate the development of another ecosystem that the climate will support and where wild processes may flourish again?

The need for new practical ethics of restoration

The rapid changes to both the global environment and our understanding of ecological restoration have resulted in a need for practical ethical guidance about when it is appropriate to restore an ecosystem and how best to perform restorations. Many restorationists have been confronted with ethical dilemmas in their work and desire advice on how to solve those dilemmas (Carpenter *et al.* 2006; Dickinson *et al.* 2006; Vidra 2006). Where should they turn for such advice? The field of environmental ethics is a growing branch of ethical inquiry and would seem to be an obvious place to turn for guidance. Unfortunately, as noted environmental ethicist Andrew Light has confessed, environmental ethicists seldom produce anything that is of direct practical use to environmentalists, ecosystem managers, planners, restorationists or even academics. Many of the conversations within the field are concerned with determining what is meant by wilderness or natural or whether particular entities have intrinsic value (Light 2006). I have read the journal *Environmental Ethics* for many years and even attended a joint conference of the International Association of Environmental Philosophy and the International Society for Environmental Ethics. The ethicists seemed to be most interested in high-level discussions of what types of ethics to apply and whether particular positions in environmental ethics fit within particular philosophical traditions. They debated whether the pragmatic ethics of John Dewey might fit a particular position and whether another might be shoehorned into the phenomenological school of Edmund Husserl. Very interesting discussions I'm sure, but of absolutely no use in trying to decide whether it is ethical to remove an invasive species from a restoration.

Restorationists are usually looking for advice on far more down-to-earth questions. How do they deal with issues of conflicts of interest? What should they do if a client wants to conduct a project in a manner that the restorationists feel will be inadequate and likely to fail? This is an especially acute issue when the restorationists have to support themselves and families and their jobs depend on working with the client. When is it permissible to use non-native, potentially invasive species in a restoration?

When it is permissible to kill large numbers of organisms in order to restore a system? How should we proceed when various stakeholder groups disagree about what they think is best for a particular system? These are not easy questions to answer and a combination of a set of guidelines and a set of values accepted within the restoration community would help to provide answers (Carpenter *et al.* 2006; Dickinson *et al.* 2006; Vidra 2006). The SER is currently in the process of producing a set of ethical guidelines for society members to adopt. The Society for Conservation Biology (SCB) did so for their society members a few years ago. The SCB code of ethics arose from a recognition that conservationists want to do the right thing, but sometimes, perhaps even often, there are conflicting right things to do (Beier 2005).

Restorationists often encounter situations where it is not immediately obvious what the best course of action is. One of the classic examples of conflicting choices occurred at Montrose Point in Lincoln Park in Chicago. Montrose Point is a strange place ecologically. The point was created from landfill dumped into Lake Michigan and was originally planted in 1929 as a landscape park in the "prairie style" favored by noted landscape architect Jens Jensen. During and after World War II the point was neglected and, among other things, a non-native honeysuckle hedge was planted along a fencerow to screen a military installation from the rest of the park area. That hedge grew up to be very large and became extremely attractive to migrating birds, so much so that area birders referred to it as the "Magic Hedge" because it attracted so many rare species of migrating birds. During the 1990s the Chicago Park District decided to change the management of the park, to remove non-native species and establish native vegetation in a way that would recreate typical Midwestern oak savanna. Many conflicts arose – growing native plants seemed like the right thing to do, preserving a valuable center for migrating bird diversity also seemed like the right thing to do, preserving park space used by many people from the city of Chicago was also desirable. All were right in their own way, but some of the plans were clearly in conflict. After a careful period of review and consultation a compromise was reached that contained many desirable features (native vegetation and savanna in some areas, retaining most of the Magic Hedge, allowing continuing human use of much of the beachfront) but no single plan or set of values was completely adopted (Gobster 2001). Individuals and groups have to make their own decisions about the best course of action and how to apply a code of ethics, but having an accepted community standard makes it easier to make decisions.

Acknowledging our limits

The development of new practical ethics within the renewed restoration program should lead to greater understanding of and a willingness to acknowledge our limitations as restorationists. When I am teaching I often use the old Clint Eastwood line from one of the Dirty Harry movies: "A man's got to know his limitations." Of course that applies to everyone – man or woman – we all need to know our limitations. There are things we can do and things we can't do. There are things we know and things we don't know. Even more important, there are things that we don't know we don't know – unknown unknowns. And we have to acknowledge that those are out there and that unknown unknowns always have the potential to render our best laid plans ineffective or even wrong. One of my goals as a restorationist is that my choices and work should not make the situation worse. Unfortunately, there is sometimes no way of knowing in advance whether a particular manipulation will cause unforeseen consequences that ultimately harm the ecosystem.

Macquarie Island is a World Heritage Site, preserved because it is home to a unique, endemic tundra-like flora and also because it is the roosting site for many seabirds. Unfortunately, rabbits were introduced to Macquarie in 1878. The rabbits prospered and by the 1960s had reached such high abundance that they were destroying the native vegetation. A strict program of rabbit control (in particular the use of the *Myxoma* virus to kill rabbits) was undertaken in 1968 and within ten years the rabbit population was reduced to the point that native vegetation began to recover. Along with rabbits, Macquarie had a large population of feral cats that were also introduced to the island in the nineteenth century. The cats ate at least a few rabbits and when the rabbit population declined, the cats switched to preying on nesting seabirds, which had a devastating impact on seabird populations. So in 1985 a cat eradication program was begun and by 2000 there were no cats left on Macquarie. The effectiveness of *Myxoma* declined and the rabbit population, now released from predation by cats, rapidly increased, with severely negative consequences for the native vegetation (Bergstrom *et al.* 2009). Here the seemingly right action – removing cats to protect seabirds – had unintended consequences that led to a rabbit population increase and increased destruction of a unique flora. Any management action we undertake has the potential to unleash unintended effects. Acknowledging our limitations and recognizing that all systems have unknown unknowns should lead us to act with humility and respect for the systems we are restoring and managing.

Yet we live in a world in which we must act. Ecological restoration, like conservation biology, is a crisis discipline in which we have to act without having perfect knowledge and thus without perfect ability to predict the outcomes of our actions (Soulé 1986). So how do we proceed in a world in which restoration is absolutely essential in order to maintain biodiversity, ecosystem functions and repair the broken connections between nature and culture? To paraphrase Voltaire, we cannot let the search for perfect solutions be the enemy of making good choices and carrying out good restorations. Our knowledge and skills are incomplete and are constantly being improved. Ideally we want to leave behind restorations that preserve the parts and processes typical of each ecosystem so that future restorationists can correct any mistakes we might make. That is good restoration. Attempting to do restorations that have no mistakes at all is a search for perfection that would paralyze us into inaction.

We will be well served by adopting the fundamental principle of medical ethics – *primum non nocere* – first do no harm. Restorationists, like physicians, are faced with a situation in which the patient (for us an ecosystem) is suffering some loss of function. The patient's structure may have been damaged, thus causing the loss of function. Sometimes the cause is obvious and we have techniques that will allow us to help the patient recover full function. But sometimes, probably more frequently, the cause is not obvious. There may be multiple causes. There are many procedures we might use either alone or in concert to assist recovery. In those cases we have to use methods that will not lead to additional harm – or perhaps as the example from Macquarie Island shows us – at the least we must use methods that will not knowingly cause additional harm. We must remember that ultimately the patient (ecosystem) will heal itself – it must use its own physiology to resolve and recover from illness (Clewell and Aronson 2007). All we can do is initiate the conditions that will facilitate healing and then step back to monitor the situation and provide additional assistance if necessary. And, following Saint Aldo's advice on intelligent tinkering, we must save all the parts, because even if we don't know what they all do, they may all be important in the functioning of our ecosystem (Leopold 1949). Above all we must be cautious in our choices and not so eager that we immediately start manipulating damaged ecosystems before we have carefully examined their past history, current conditions, the potential causes for their condition and what we have learned from other restorations that may apply to new restorations.

Acknowledging limitations, actively seeking to do no harm and exercising caution will slow us down as we plan and carry out restorations. I, like many people in the field, feel a pressing need to get on with the work of restoration, but rushing headlong into restoration projects will do us and the environment no good. Slowing down long enough to consider the full weight of our actions will help us avoid mistakes. And perhaps even more importantly, it may help us ward off hubris. Slowing down should prevent us from feeling that we have the technical abilities to solve any environmental problem regardless of how large or complex – a feeling which could lead to us continuing our destructive patterns of land use because we might mistakenly think, yes, we can fix this later.

Steps to take to accomplish renewed restoration

So, if restorationists must exercise caution while acknowledging our limitations, what can we do to accomplish the renewed restoration that is necessary in the twenty-first century? There are many things we can do. I will list what I think are the six most important steps for restorationists in the next 100 years. Other restorationists may have other priorities, but for me these six are essential.

1 We must be able to document and demonstrate the benefits of already-existing restorations. And we must make sure that future restorations include an explicit statement of goals and are designed with monitoring protocols that will allow us to determine how effective the restorations are. Unfortunately, in the past, restorationists have not been as rigorous as they should be at clearly stating goals and setting up monitoring programs. For example, in a database of 37,000 stream restoration projects in the United States, only 10 percent report monitoring stream quality after the restoration. Of those 37,000 projects, 20 percent did not list any goals in their report and 40 percent provided no data on the cost of the restoration (Fleishman *et al.* 2011).

   Many people in the restoration community have stressed the need to document economic benefits of restoration and would like us to focus on the ecosystem services provided by restored ecosystems (Aronson *et al.* 2007, 2010). But we have not even done a good job of reporting benefits to biodiversity and basic ecosystem function (Fleishman *et al.* 2011). Clearly many government organizations, NGOs and corporations that might fund restoration projects will be especially interested in the economic return provided by restoration. But economic benefits are only a small part of what can be achieved by good restoration. Benefits to the entire environment are best documented by measuring biodiversity and ecosystem function. We also need to report the cultural benefits of ecological restoration, many of which will not be amenable to direct economic evaluation. If we focus exclusively on the economic return and ecosystem services provided by ecological restoration, we risk going down a path along which restoration becomes a highly technical engineering exercise conducted only by trained professionals (Higgs 2005). The full benefits of ecological restoration will only be realized if all people have the opportunity to participate in the process and thus restore their own relationship to the ecological world.

2 We must restore as many ecosystems, over as large an area as possible. We must also preserve as much remaining wild and semi-wild area as possible. At the current time only about 5 percent of the land is set aside in nature preserves and considerably less than that is available for ecological restoration. Some people fear we will not be able to add much more land to the amount already in preserves (Miller 2005). But we know that more area of restored and preserved ecosystems will allow us to preserve more biodiversity and will provide better ecosystem function and services (Millennium Ecosystem Assessment 2005).

It is especially important that we preserve and restore ecosystems in order to provide space for species to move to as local and global climates change. Even small increases in the amount of available habitat result in increased probability of species being able to disperse and expand their range. Increases in available habitat also result in more rapid dispersal. A recent study found that the pattern of habitat restoration was not important. Restorations scattered randomly, evenly or at sites especially selected by stakeholders all promoted species dispersal as long as the total area of available habitat increased (Hodgson *et al.* 2011). This is a hopeful result because it shows that we don't have to worry about finding the perfect spacing for restorations in order to provide benefits for many species – we merely have to increase the area that is suitable for them, which greatly simplifies the planning process.

3 We are almost certainly going to have to actively move individuals from one location to another in order to establish new populations. This movement of individuals from one place to another, often referred to as managed relocation or translocation, is one of the most active forms of manipulation we can perform during ecological management and therefore is also one of the most controversial actions we can perform.

Some translocations, such as moving individuals into an area where a species occurred historically but where for some reason its population went extinct, are usually seen as fairly straightforward and therefore acceptable practice. However, reintroductions to the historical range do have some potential problems. Why did the population go extinct and has the cause of extinction been removed? Sometimes historical records are not as accurate as we would like so there can be questions of whether the species ever occurred at the proposed reintroduction site. Sometimes a translocation may occur for conservation and biodiversity reasons – moving individuals to an area where the species never occurred in the past, but which may now serve as a safe site that has the proper climate for the species or where predators and diseases are absent. Establishing conservation populations on islands that were never home to the species has been used successfully in New Zealand as a way to preserve biodiversity and provide a refuge for vulnerable species. Species may also be moved to new areas as part of ecological engineering programs in which new ecosystems are created from novel combinations of species that will hopefully develop into ecosystems that are sustainable and provide cultural and ecological benefits even in the face of continued environmental change. Such engineering translocations are the most controversial of all (Seddon 2010).

Although I think many restoration projects will include translocations in the twenty-first century, these are manipulations that require extra caution on our part, especially when moving individuals out of their species' historical ranges. Given our excellent abilities to move species around – sometimes to the detriment of the environment and ourselves – translocations should be a method of last resort. If we can demonstrate that a species is in terminal decline and that it cannot survive in its current position, then we should move individuals to some place better for them. But if they are still doing well in their current home, then we should refrain from translocating that species.

4 We need to make more effective use of urban and other highly domesticated habitats in order to promote biodiversity, ecosystem function and ecosystem services. Traditionally, restorationists have focused on restoring ecosystems that are located in rural and semi-wild habitats. But given that only about 5 percent of the earth's land surface is currently in preserves and that the growing human footprint results in increasingly fragmented landscapes, we need to restore biodiversity and ecosystem function wherever we can.

As long as there are people on the earth we cannot possibly restore all the earth to wild conditions. Because the human population is predicted to increase in size at least until 2050, we must find ways to increase biodiversity and ecosystem function in human-dominated habitats as well as in semi-wild habitats. This is not a new idea. Aldo Leopold felt that the ideal conservationist was a farmer who found room on the farm for both the domesticated species with which the farmer made a living and for wild species of no direct benefit to the farmer (Leopold 1939).

Every farm should have room in corners of fields and pastures, under the eaves of sheds and barns, where many species could find refuge and a way to make a living. Urban landscapes contain many small private gardens which represent an excellent resource that could help preserve biodiversity. Gardens occupy anywhere from 16 to 36 percent of urban area (depending on the city), so there is a significant amount of space for gardeners and restorationists to work with. Of course, urban gardens are mostly privately owned by many different individuals who may have divergent ideas about what makes a good garden, but not all gardeners have to be doing the same thing to achieve ecological benefits. Encouraging gardeners to use more native species, to plant a large diversity of plants and to minimize or eliminate the use of pesticides can have a tremendous benefit when individual actions are multiplied over hundreds to thousands of urban gardens. Coordinated changes to garden practices can be achieved either via top-down controls (regulations and financial incentives) or bottom-up programs such as educational outreach and community-based initiatives (Goddard *et al.* 2010). In all cases, with today's increasingly urbanized human population, the first and primary place where many people will be able to develop a personal relationship with the environment will be in the domesticated landscape of public and private lawns and gardens. Restoration of the relationship between culture and nature will be greatly enhanced by recognizing the importance of urban and domesticated habitats in ecological restoration.

5 Good ecological restoration depends on the use of adaptive management. The implementation, continuing monitoring and adjusting of management plans will be vitally important due to the projected pace and magnitude of environmental change in the twenty-first century. There is broad global recognition of the need to maintain biodiversity. The United Nations Convention on Biological Diversity (CBD) has 193 signatory parties. In 2002 those parties agreed to take actions that would significantly reduce the rate of decline of biodiversity. And yet in 2010, biodiversity was continuing to decline at the same rate as it was in 2002. Policies that might reduce the rate of biodiversity loss have either not been adequate or have not been fully implemented – frequently it is the latter (Rands *et al.* 2010). Most environmental management plans were developed prior to 2000, prior to our recognition of the increasing pace and great magnitude of environmental change (Lawler *et al.* 2010). Those plans are almost certainly inadequate to allow effective management of our diminishing biodiversity and damaged habitats in the twenty-first century. As noted throughout this book, there are many uncertainties about the exact magnitude and effects of projected climate change for any particular location. Therefore our management plans have to be flexible. Ecological restoration plans must be designed in order to achieve a goal of authentic restorations – the restored ecosystem must have authentic form and function. Our prescription for authenticity must be adaptive so that we can adjust our management options to allow for the use of new or different techniques and flexibility in species composition as the ecosystem changes over time (Lawler *et al.* 2010; Altman *et al.* 2011).

Ultimately, restorationists, managers, policy-makers and stakeholders must work together to ensure everyone understands the adaptive process, recognizes the environmental changes affecting the ecosystem and is at least comfortable with how management plans are implemented.

6 Finally, restorationists must become effective advocates for ecological restoration. On one hand it should be easy to present a compelling case for the benefits of ecological restoration – the value (ecological, cultural and economic) of maintaining global biodiversity and improving local ecosystem function and services. On the other hand, many restoration ecologists are uncomfortable in the role of advocate. Restorationists who come from non-scientific backgrounds, and especially those who are volunteers who became involved in restoration from a desire to do good for the earth, may be perfectly comfortable with

advocacy. However, scientists are often trained to be as objective as possible, to let the facts speak for themselves and to avoid entering the political arena because political activity may cause a loss of scientific credibility. But everyone involved with restoration must learn to be an effective advocate.

When scientists provide an assessment of the data and also provide their own opinions and policy recommendations they can be especially effective advocates and can help improve policy- and decision-making. Anyone acting as an advocate for restoration must be sure to present the data as completely and accurately as possible – selectively citing helpful data (in terms of promoting a particular position) and ignoring other data will hurt the credibility of everyone involved with restoration.

Educational outreach is an absolutely essential part of advocacy. The general public and policy-makers must have access to the latest and best scientific information. For scientists this means discussing the results of scientific research in a much wider forum than just peer-reviewed journals – most people will never read those journals. We have to reach out to the news media and make contacts to help spread the word about restoration. The use of blogging and social media will be an increasingly important way to reach the public. As a part of our advocacy we must make sure that not only do policy-makers have access to the best information, we must also make sure they actually review that information before making decisions. Because restoration is a crisis discipline and we often have to make recommendations before we have perfect data, we will often be making recommendations based on our gut reaction. There is nothing wrong with that, but we have to be clear about when a recommendation is based on well-established data and when it is a hunch. Finally, we can do advocacy work at three levels – locally, regionally and nationally/internationally. Each restorationist will have to decide the level at which they are most comfortable working, but each of us must be willing to do some advocacy in order to achieve the renewed restoration necessary in the next 100 years.

Reasons to be cheerful

We have a difficult balancing act as we work to achieve renewed restoration. Restorationists must balance the need to develop and use the best scientific information with the need to be effective advocates for restoration. We must carry out restorations that satisfy both ecological and cultural needs. We must approach restoration with caution and humility, but we must also act to restore as much area as rapidly as possible in order to preserve significant amounts of biodiversity and ecosystem function.

In many ways our work as restorationists will have to follow the Goldilocks principle – not too much of one thing (not too rigid an adherence to historical ecological conditions; porridge too hot) and not too little of something (too little consideration of historical fidelity, too little concern about our limitations; porridge too cold) – in order to arrive at the just-right solution that works best for humans and the entire natural world. Perhaps the Goldilocks principle sounds simplistic, but carefully considered choices seeking the just-right solution will be more likely to be wise choices and will result in more and better connections between ecological needs and function and human needs and values. Good restoration that creates authentic ecosystems will only happen when we are fully engaged with the process and carefully observing how the ecosystem responds to our management choices and methods. If we listen closely enough to what the ecosystem is telling us and to how we all respond to that message, then we can create ecosystems with ecological integrity and renew the cultural–natural connection.

There are many reasons to be optimistic about the future of ecological restoration, despite the challenges in front of us. When I was a child growing up in the American Midwest, native prairie plants were considered to be unwanted weeds, but today they are valued as beautiful and necessary parts of the landscape. There has been a sea change in attitude toward an entire ecosystem in a relatively short time. When people have good models to follow, they may swiftly change attitudes and practices (Goddard *et al.* 2010). All it takes is a few committed individuals who can show how to do something and many people become excited and quickly adopt restoration practices. Just a little bit of exposure to the beauty of biodiversity and the joys of restoration work creates dedicated restorationists in even the most highly urbanized, synthetic environments such as the middle of Liverpool (Scott and Whitbread-Abrutat 2010). People have tremendous capacity to renew themselves and their environment in highly beneficial ways; we simply have to reach out and help them realize their potential to become part of the necessary renewal.

There is much work that must be done to solve the many environmental challenges facing us in the twenty-first century. Ecological restoration will be one of our most effective tools for this work. We can do the work and by engaging more and more people in the effort we will do the work. I am confident that the twenty-first century will be a century of positive renewal if we apply our will and make the effort – it is the only sensible choice before us.

References

Adams, D. and Carwardine, M. (1992) *Last Chance to See*, Ballantine Books, London.

Agrawal, A. and Ostrom, E. (2006) Political Science and conservation biology: a dialog of the deaf, *Conservation Biology, 20*(3), 681–682.

Allison, S.K. (2002) When is a restoration successful? Results from a 45-year-old tallgrass prairie restoration, *Ecological Restoration, 20*, 10–17.

Allison, S.K. (2004) What do we mean when we talk about ecological restoration? *Ecological Restoration, 22*(4), 281–286.

Allison, S.K. (2011) The paradox of invasive species: do restorationists worry about them too much or too little? In I.D. Rotherham and R. Lambert (eds.) *Invasive and Introduced Plants and Animals: Human Perceptions, Attitudes, and Approaches to Management*, Earthscan Press, London, pp. 265–275.

Altman, I., Blakeslee, A.M.H., Osio, G.C., Rillahan, C.B., Teck, S.J., Meyer, J.J., Byers, J.E. and Rosenberg, A.A. (2011) A practical approach to implementation of ecosystem-based management: a case study using the Gulf of Maine marine ecosystem, *Frontiers in Ecology and the Environment, 9*(3), 183–189.

Anderson, M.K. (2005) *Tending the Wild: Native American Knowledge and the Management of California's Natural Resources*, University of California Press, Berkeley, CA.

Anderson, R.A. (2006) Evolution and origin of the central grassland of North America: climate, fire and mammalian grazers, *Journal of the Torrey Botanical Society, 133*, 626–647.

Aronson, J., Milton, S.J. and Blignaut, J.N. (eds.) (2007) *Restoring Natural Capital: Science, Business and Practice*, Island Press, Washington, DC.

Aronson, J., Blignaut, J.N., Milton, S.J., Le Maitre, D., Esler, K.J., Limouzin, A., Fontaine, C., De Wit, M.P., Mugido, W., Prinsloo, P., van der Elst, L. and Lederer, N. (2010) Are socioeconomic benefits of restoration adequately quantified? A meta-analysis of recent papers (2000–2008) in *Restoration Ecology* and 12 other scientific journals, *Restoration Ecology, 18*(2), 143–154.

Bakker, J.P. and Berendse, F. (1999) Constraints in the restoration of ecological diversity in grassland and heathland communities. *Trends in Ecology & Evolution, 14*(2), 63–68.

Baldwin, A.D., DeLuce, J. and Pletsch, C. (eds.) (1994) *Beyond Preservation: Restoring and Inventing Landscapes*, University of Minnesota Press, Minneapolis, MN.

Ball, L. (2008) Called to action: environmental restoration by artists, *Ecological Restoration, 26*(1), 27–32.

Barnosky, A.D., Matzke, N., Tomiya, S., Wogan, G.O.U., Swartz, B., Quental, T.B., Marshall, C., McGuire, J.L., Lindsey, E.L., Maguire, K.C., Mersey, B. and Ferrer, E.A. (2011) Has the Earth's sixth mass extinction already arrived? *Nature, 471*(7336), 51–57.

BBC (2011) Llamas move fish to cooler waters in Lake District. www.bbc.co.uk/news/uk-england-cumbria-13042511, retrieved April 12, 2011.

Beier, P. (2005) Being ethical as conservation biologists and as a society, *Conservation Biology, 19*(1), 1–2.

Bergstrom, D.M., Lucieer, A., Kiefer, K., Wasley, J., Belbin, L., Pedersen, T.K. and Chown, S.L. (2009) Indirect effects of invasive species removal devastate World Heritage Island, *Journal of Applied Ecology, 46*(1), 73–81.

Berkes, F., Colding, J. and Folke, C. (2000) Rediscovery of traditional ecological knowledge as adaptive management, *Ecological Applications, 10*(5), 1251–1262.

Botkin, D.B. (1990) *Discordant Harmonies: A New Ecology for the 21st Century*, Oxford University Press, New York, NY.

Bowers, K. (2006) Human systems, *Ecological Restoration, 24*(2), 70.

Box, J. (1996) Setting objectives and defining outputs for ecological restoration and habitat creation. *Restoration Ecology, 4*(4), 427–432.

Bradley, B.A. and Wilcove, D.S. (2009) When invasive plants disappear: transformative restoration possibilities in the western United States resulting from climate change, *Restoration Ecology, 17*(5), 715–721.

Bradley, B.A., Blumenthal, D.M., Wilcove, D.S. and Ziska, L.H. (2010) Predicting plant invasions in an era of global change, *Trends in Ecology and Evolution, 25*(5), 310–318.

Bradshaw, A.D. (1984) Ecological principles and land reclamation practice, *Landscape Planning, 11*, 35–48.

Bradshaw, A.D. (1987) Restoration: an acid test for ecology. In W.R. Jordan III, M.E. Gilpin and J.D. Aber (eds.), *Restoration Ecology: A Synthetic Approach to Ecological Research*, Cambridge University Press, Cambridge, pp. 23–29.

Bradshaw, A.D. (2002) Introduction and philosophy. In M.R. Perrow and A.J. Davy, (eds.), *Handbook of Ecological Restoration: Vol. 1 Principles of Restoration*, Cambridge University Press, Cambridge, UK, pp. 3–9.

Bradshaw, A.D. and Chadwick, M.J. (1980) *The Restoration of Land: The Ecology and Reclamation of Derelict and Degraded Land*, University of California Press, Berkeley, CA.

Brudvig, L.A. (2011) The restoration of biodiversity: where has research been and where does it need to go? *American Journal of Botany, 98*, 549–558.

Brunson, M.W. (2000) Managing naturalness as a continuum: setting limits of acceptable change. In P.H. Gobster and B. Hull (eds.), *Restoring Nature: Perspectives from the Social Sciences and Humanities*, Island Press, Washington. DC, pp. 229–244.

Cabin, R.J. (2007) Science-driven restoration: a square grid on a round earth?, *Restoration Ecology, 15*(1), 1–7.

Cabin, R.J. (2011) *Intelligent Tinkering: Bridging the Gap Between Science and Practice*, Island Press, Washington, DC.

Cabin, R.J., Clewell, A., Ingram, M., McDonald, T. and Temperton, V. (2010) Bridging restoration science and practice: results and analysis of a survey from the 2009 Society for Ecological Restoration international meeting. *Restoration Ecology, 18*(6), 783–788.

Cairns, J.J. (1980) *The Recovery Process in Damaged Ecosystems*, Ann Arbor Science Publications, Ann Arbor, MI.

Callicott, J.B. (1990) Whither conservation ethics? *Conservation Biology, 4*, 15–20.

Cardille, J.A. and Lambois, M. (2010) From the redwood forest to the gulf stream waters: human signature nearly ubiquitous in representative US landscapes, *Frontiers in Ecology and the Environment, 8*(3), 130–134.

Carpenter, A., Finley, E., Gao, Y., Lin, C., Nuding, A., Shaheen, P., Stewart, L., Sun, X., Taranto, M., Tilley, A., Waggoner, L., Xu, H. and Vidra, R.L. (2006) Developing a code of ethics for restorationists, *Ecological Restoration, 24*(2), 105–108.

Carson, R. (1962) *Silent Spring*. Fawcett Crest, Greenwich, CT.

Chazdon, R.L. (2008) Beyond deforestation: restoring forests and ecosystem services on degraded lands. *Science, 320*(5882), 1458–1460.

Cheney, J. and Weston, A. (1999) Environmental ethics as environmental etiquette: toward an ethics-based epistemology. *Environmental Ethics, 21*(2), 115–134.

Chestney, N. (2009, November 26) Forest area bigger than Canada can be restored. *Thomson Reuters*. www.reuters.com/article/idUSTRE5AP3AG20091126.

Choi, Y.D. (2004) Theories for ecological restoration in changing environment: toward "futuristic" restoration, *Ecological Research, 19*, 75–81.

Clark, J.A. and May, R.M. (2002) Taxonomic bias in conservation research, *Science, 297*(5579), 191–192.

Clewell, A.F. and Aronson, J. (2006) Motivations for the restoration of ecosystems. *Conservation Biology, 20*(2), 420–428.

Clewell, A.F. and Aronson, J. (2007) *Ecological Restoration: Principles, Values and Structure of an Emerging Profession*, Island Press, Washington, DC.

Cline, A. (1994) The little hut on the prairie: the ritual uses of restoration. In A.D. Baldwin, J. de Luce and C. Pletsch (eds.), *Beyond Preservation: Restoring and Inventing Landscapes*, University of Minnesota Press, Minneapolis, MN, pp. 216–225.

Cliquet, A., Backes, C., Harris, J. and Howsam, P. (2009) Adaptation to climate change: legal challenges for protected areas. *Utrecht Law Review, 5*(1), 158–175.

Coates, P. (2007) *American Perceptions of Immigrant and Invasive Species: Strangers on the Land*, University of California Press, Berkeley, CA.

Cole, D.N., Higgs, E. and White, P.S. (2010) Historical fidelity: maintaining legacy and connection to heritage. In D.N. Cole and L. Yung (eds.), *Beyond Naturalness: Rethinking Park and Wilderness Stewardship in an Era of Rapid Change*, Island Press, Washington, DC, pp. 125–141.

Collins, S.L. and Wallace, L.L. (eds.) (1990) *Fire in North American Tallgrass Prairies*, University of Oklahoma Press, Norman, OK.

Comin, F.A. (2010) The challenges of humanity in the twenty-first century and the role of ecological restoration. In F.A. Comin (ed.), *Ecological Restoration: A Global Challenge*, Cambridge University Press, Cambridge, UK, pp. 3–20.

Costanza, R. (2010) The value of a restored earth and its contribution to a sustainable and desirable future. In F.A. Comin (ed.), *Ecological Restoration: A Global Challenge*, Cambridge University Press, Cambridge, UK, pp. 78–90.

Cronon, W. (1995) The trouble with wilderness; or getting back to the wrong nature. In W. Cronon (ed.), *Uncommon Ground: Rethinking the Human Place in Nature*, W.W. Norton Co, New York, NY, pp. 69–90.

Cronon, W. (2003) The riddle of the Apostle Islands: how do you manage a wilderness full of human stories? *Orion*, May/June, 36–42.

Crutzen, P.J. and Stoermer, E.F. (2000) The "Anthropocene," *Global Change Newsletter, 41*, 17–18.

Curtis, D.J. (2009) Creating inspiration: the role of the arts in creating empathy for ecological restoration, *Ecological Management and Restoration, 10*(3), 174–184.

Curtis, J.T. and Partch, M.L. (1948) Effect of fire on the competition between blue grass and certain prairie species, *The American Midland Naturalist, 39*, 437–443.

Czech, B. (2004) A chronological frame of reference for ecological integrity and natural conditions. *Natural Resources Journal, 44*(4), 1113–1136.

Daily, G.C., Polasky, S., Goldstein, J., Kareiva, P.M., Mooney, H.A., Pejchar, L., Ricketts, T.H., Salzman, J. and Shallenberger, R. (2009) Ecosystem services in decision making: time to deliver. *Frontiers in Ecology and the Environment, 7*(1), 21–28.

Danielsen, F., Skutsch, M., Burgess, N.D., Jensen, P.M., Andrianandrasana, H., Karky, B., Lewis, R., Lovett, J.C., Massao, J., Ngaga, Y., Phartiyal, P., Poulsen, M.K., Singh, S.P., Solis, S., Sørensen, M., Tewari, A., Young, R. and Zahabu, E. (2011) At the heart of REDD+: a role for local people in monitoring forests? *Conservation Letters, 4*(2), 158–167.

Davis, M.A. and Slobodkin, L.B. (2004) The science and values of restoration ecology, *Restoration Ecology, 12*, 1–3.

Davis, M.B. (1987) Invasions of forest communities during the Holocene: beech and hemlock in the Great Lakes region, *Colonization, Succession and Stability: 26th BES Symposium*, Blackwell Science Inc., Oxford, pp. 373–393.

Davis, W. (1996) *One River: Explorations and Discoveries in the Amazon Rain Forest*, Touchstone, New York, NY.

DeClerck, F.J.A., Chazdon, R., Holl, K.D., Midler, J.C., Finegan, B., Martinez-Salinas, A., Imbach, P., Canet, L. and Ramos, Z. (2010) Biodiversity conservation in human-modified landscapes of Mesoamerica: past, present and future. *Biological Conservation, 143*, 2301–2313.

DeGroot, R.S., Wilson, M.A. and Boumans, R.M.J. (2002) A typology for the classification, description and valuation of ecosystem functions, goods and services. *Ecological Economics, 41*(3), 393–408.

DeLestard, L.P.G. (1967) *A History of the Sudbury Forest District, District History Series Number 21*, Ontario Department of Lands and Forests, Toronto, ON.

Denevan, W.M. (1992) The pristine myth: the landscape of the Americas in 1492, *Annals of the Association of American Geographers, 82*(3), 369–385.

Dickinson, W., Ferreyra, J., Imbesi, K.L., Joshi, S., Kingsolver, C., Klein, E., Lessios, N., Ng, A., Stamp, T., White, K., Xu, D. and Vidra, R.L. (2006) The ethical challenges faced by ecological restorationists, *Ecological Restoration, 24*(2), 102–104.

Dobson, A.P., Bradshaw, A.D. and Baker, A.J.M. (1997) Hopes for the future: restoration ecology and conservation biology. *Science, 277*(5325), 515–522.

Donlan, C.J., Berger, J., Bock, C.E., Bock, J.H., Burney, D.A., Estes, J.A., Foreman, D., Martin, P.S., Roemer, G.W., Smith, F.A., Soulé, M.E. and Greene, H.W. (2006) Pleistocene rewilding: an optimistic agenda for twenty-first century conservation, *American Naturalist, 168*(5), 660–681.

Dorsey, J.W. (2003) Brownfields and greenfields: the intersection of sustainable development and environmental stewardship. *Environmental Practice, 5*(1), 69–76.

Doyle, M. (2008) Conclusion: assessing ecosystem restoration projects. In M. Doyle and C.A. Drew (eds.), *Large Scale Ecosystem Restoration: Five Case Studies from the United States,* Island Press, Washington, DC, pp. 291–299.

Dudley, N. (2011) *Authenticity in Nature: Making Choices About Naturalness in Ecosystems,* Earthscan Books, Abingdon, UK.

Dudley, N., Mansourian, S. and Vallauri, D. (2005) Forest landscape restoration in context. In S. Mansourian, D. Vallauri and N. Dudley (eds.), *Forest Restoration in Landscapes: Beyond Planting Trees,* Springer, New York, NY, pp. 3–7.

Duffy, J.E. (2009) Why biodiversity is important to the functioning of real-world ecosystems, *Frontiers in Ecology and the Environment, 7*(8), 437–444.

Duncan, R.P. and Williams, P.A. (2002) Ecology: Darwin's naturalization hypothesis challenged, *Nature, 417*(6889), 608–609.

Dunwiddie, P.W. (1992) On setting goals: from snapshots to movies and beyond..., *Restoration and Management Notes, 10*(2), 116–119.

Egan, D. (1990) Historic initiatives in ecological restoration, *Restoration and Management Notes, 8,* 83–90.

Ehrenfeld, J.G. (2000) Defining the limits of restoration: the need for realistic goals, *Restoration Ecology, 8*(1), 2–9.

Ehrenfeld, J.G. and Toth, L.A. (1997) Restoration ecology and the ecosystem perspective, *Restoration Ecology, 5,* 307–317.

Eisenberg, E. (1998) *The Ecology of Eden,* Vintage Books, New York, NY.

Elliott, R. (1982) Faking nature, *Inquiry, 25,* 81–93.

Ellis, E.C. and Ramankutty, N. (2008) Putting people in the map: anthropogenic biomes of the world, *Frontiers in Ecology and the Environment, 6*(8), 439–447.

Elton, C.S. (1958) *The Ecology of Invasions by Animals and Plants,* University of Chicago Press, Chicago, IL.

Ewel, J.J. and Putz, F.E. (2004) A place for alien species in ecosystem restoration, *Frontiers in Ecology and the Environment, 2*(7), 354–360.

Farber, S.C., Costanza, R. and Wilson, M.A. (2002) Economic and ecological concepts for valuing ecosystem services, *Ecological Economics, 41*(3), 375–392.

Fischer, A. and Fischer, H. (2006) Restoration of forests. In J. van Andel and J. Aronson (eds.), *Restoration Ecology: The New Frontier,* Blackwell Publishing, Malden, MA, pp. 124–140.

Fischer, J. and Lindenmayer, D.B. (2007) Landscape modification and habitat fragmentation: a synthesis, *Global Ecology and Biogeography, 16,* 265–280.

Fleishman, E., Blockstein, D.E., Hall, J.A., Mascia, M.B., Rudd, M.A., Scott, J.M., Sutherland, W.J., Bartuska, A.M., Brown, A.G., Christen, C.A., Clement, J.P., Dellasala, D., Duke, C.S., Eaton, M., Fiske, S.J., Gosnell, H., Haney, J.C., Hutchins, M., Klein, M.L. and Marqusee, J. (2011) Top 40 priorities for science to inform US conservation and management policy, *Bioscience, 61*(4), 290–300.

Flynn, D.F.B., Mirotchnick, N., Jain, M., Palmer, M.I. and Naeem, S. (2011) Functional and phylogenetic diversity as predictors of biodiversity–ecosystem-function relationships, *Ecology, 92*(8), 1573–1581.

Folke, C., Hahn, T., Olsson, P. and Norberg, J. (2005) Adaptive governance of social–ecological systems. *Annual Review of Environment and Resources, 30,* 441–473.

Ford, J. and Martinez, D. (2000) Traditional ecological knowledge, ecosystem science, and environmental management, *Ecological Applications, 10*(5), 1249–1250.

Fox, D. (2007) Ecology: back to the no-analog future? *Science, 316*(5826), 823–825.

Fulghum, R. (1988) *All I Really Need to Know I Learned in Kindergarten: Uncommon Thoughts on Common Things*, Villard Books, New York, NY.

Galatowitsch, S.M. (2009) Carbon offsets as ecological restorations. *Restoration Ecology, 17*(5), 563–570.

Gao, Q., Li, Y., Wan, Y., Jiangcun, W., Qin, X. and Wang, B. (2009) Significant achievements in protection and restoration of alpine grassland ecosystem in northern Tibet, China, *Restoration Ecology, 17*(3), 320–323.

Gladwell, M. (2000) *The Tipping Point: How Little Things Can Make a Big Difference*, Little Brown and Co., New York, NY.

Gobster, P.H. (2001) Visions of nature: conflict and compatibility in urban park restoration, *Landscape and Urban Planning, 56*(1–2), 35–51.

Gobster, P.H. and B. Hull (eds.) (2000) *Restoring Nature: Perspectives From the Social Sciences and Humanities*, Island Press, Washington, DC.

Goddard, M.D., Dougill, A.J. and Benton, T.G. (2010) Scaling up from gardens: biodiversity conservation in urban environments, *Trends in Ecology & Evolution, 25*(2), 90–98.

Golubiewski, N. (2008) Species influences upon ecosystem function. In M. McGinley (ed.), *Encyclopedia of Earth*, Environmental Information Coalition, National Council for Science and the Environment, Washington, DC, pp. 1–6.

Gómez-Aparicio, L. (2009) The role of plant interactions in the restoration of degraded ecosystems: a meta-analysis across life-forms and ecosystems. *Journal of Ecology, 97*(6), 1202–1214.

Goodwin, L. (2010) One-fifth of Pakistan under water as flooding disaster continues. http://news.yahoo.com/blogs/upshot/one-fifth-pakistan-under-water.html, retrieved October 6, 2010.

Graber, D. (2003) Ecological restoration in wilderness: natural versus wild in national park service wilderness, *The George Wright Forum, 20*(3), 34–41.

Grant, C.D., Ward, S.C. and Morley, S.C. (2007) Return of ecosystem function to restored bauxite mines in western Australia, *Restoration Ecology, 15*(suppl. 4), S94–S103.

Groffman, P.M., Stylinski, C., Nisbet, M.C., Duarte, C.M., Jordan, R., Burgin, A., Andrea Previtali, M. and Cary, J.C. (2010) Restarting the conversation: challenges at the interface between ecology and society, *Frontiers in Ecology and the Environment, 8*(6), 284–291.

Gross, M. (2006) Beyond expertise: ecological science and the making of socially robust restoration strategies, *Journal for Nature Conservation, 14*, 172–179.

Hall, M. (1997) Co-workers with nature: the deeper roots of restoration, *Restoration and Management Notes, 15*, 173–178.

Hall, M. (2005) *Earth Repair: A Transatlantic History of Environmental Restoration*, University of Virginia Press, Charlottesville, VA.

Halle, S. (2007) Present state and future perspectives of restoration ecology: introduction, *Restoration Ecology, 15*(2), 304–306.

Halliday, S. (1999) *The Great Stink of London: Sir Joseph Bazalgette and the Cleansing of the Victorian Metropolis*, Sutton Press, New York, NY.

Hansen, A.J., Knight, R.L., Marzluff, J.M., Powell, S., Brown, K., Gude, P.H. and Jones, K. (2005) Effects of exurban development on biodiversity: patterns, mechanisms, and research needs, *Ecological Applications, 15*(6), 1893–1905.

Hansen, L., Hoffman, J., Drews, C. and Mielbrecht, E. (2010) Designing climate-smart conservation: guidance and case studies, *Conservation Biology, 24*(1), 63–69.

Hardin, G. (1968) The tragedy of the commons, *Science, 162*(3859), 1243–1248.

Harris, J.A. and van Diggelen, R. (2006) Ecological restoration as a project for global society. In J. van Andel and J. Aronson (eds.), *Restoration Ecology: The New Frontier*, Blackwell Publishing, Malden, MA, pp. 3–15.

Harris, J.A., Hobbs, R.J., Higgs, E. and Aronson, J. (2006) Ecological restoration and global climate change. *Restoration Ecology, 14*(2), 170–176.

Helford, R.M. (1999) Rediscovering the presettlement landscape: making the oak savanna ecosystem "real," *Science Technology and Human Values, 24*(1), 55–79.

Higgs, E. (1997) What is good ecological restoration? *Conservation Biology, 11*(2), 338–348.

Higgs, E. (2003) *Nature by Design: People, Natural Process, and Ecological Restoration*, MIT Press, Cambridge, MA.

Higgs, E. (2005) The two-culture problem: ecological restoration and the integration of knowledge, *Restoration Ecology, 13*, 159–164.

Higgs, E. (2010) Focal restoration. In F.A. Comin (ed.), *Ecological Restoration: A Global Challenge*, Cambridge University Press, Cambridge, UK, pp. 91–99.

Higgs, E. and Hobbs, R.J. (2010) Wild design: principles to guide interventions in protected areas. In D.N. Cole and L. Young (eds), *Beyond Naturalness: Rethinking Park and Wilderness Stewardship in an Era of Rapid Change*, Island Press, Washington, DC, pp. 234–251.

Hobbs, R.J. and Cramer, V.A. (2008) Restoration ecology: interventionist approaches for restoring and maintaining ecosystem function in the face of rapid environmental change. *Annual Review of Environment and Resources, 33*, 39–61.

Hobbs, R.J. and Harris, J.A. (2001) Restoration ecology: repairing the Earth's ecosystems in the new millennium, *Restoration Ecology, 9*, 239–246.

Hobbs, R.J., Cramer, V.A. and Kristjanson, L.J. (2003) What happens if we cannot fix it? Triage, palliative care and setting priorities in salinising landscapes, *Australian Journal of Botany, 51*(6), 647–653.

Hobbs, R.J., Arico, S., Aronson, J., Baron, J.S., Bridgewater, P., Cramer, V.A., Epstein, P.R., Ewel, J.J., Klink, C.A., Lugo, A.E., Norton, D., Ojima, D., Richardson, D.M., Sanderson, E.W., Valladores, F., Vila, M., Zamora, R. and Zobel, M. (2006) Novel ecosystems: theoretical and management aspects of the new ecological world order, *Global Ecology and Biogeography, 15*, 1–7.

Hobbs, R.J., Higgs, E. and Harris, J.A. (2009) Novel ecosystems: implications for conservation and restoration. *Trends in Ecology and Evolution, 24*(11), 599–605.

Hodgson, J.A., Thomas, C.D., Cinderby, S., Cambridge, H., Evans, P. and Hill, J.K. (2011) Habitat re-creation strategies for promoting adaptation of species to climate change, *Conservation Letters, 4*(4), 289–297.

Holland, K.M. (1994) Restoration rituals: transforming workday tasks into inspirational rites, *Restoration and Management Notes, 12*, 121–125.

Holling, C.S. (1978) *Adaptive Environmental Assessment and Management*, John Wiley and Sons, London.

Holling, C.S. (1979) Myths of ecological stability: resilience and the problem of failure. In C.F. Smart and W. Stanbury (eds.), *Studies in Crisis Management*, Butterworth and Co., Montreal, pp. 97–109.

Hulme, P.E. (2006) Beyond control: wider implications for the management of biological invasions. *Journal of Applied Ecology, 43*(5), 835–847.

Hunter, M.L. and Gibbs, J. (2007) *Fundamentals of Conservation Biology*, 3rd edn., Blackwell Publishing, Malden, MA.

Huston, M.A. (2005) The three phases of land-use change: implications for biodiversity, *Ecological Applications, 15*(6), 1864–1878.

Jackson, S.T. and Hobbs, R.J. (2009) Ecological restoration in the light of ecological history, *Science, 325*(5940), 567–569.

Janzen, D.H. (1974) The deflowering of Central America, *Natural History, 83*, 48–53.

Janzen, D.H. (1988) Tropical ecological and biocultural restoration, *Science, 239*(4837), 243–244.

Jeffords, M. and Post, S. (2005) Art: Emiquon Corps of Discovery – providing an aesthetic perspective of a large-scale wetland restoration (Illinois), *Ecological Restoration, 23*(3), 217–218.

Jordan, W.R., III (2003) *The Sunflower Forest: Ecological Restoration and the New Communion with Nature*, University of California Press, Berkeley, CA.

Kareiva, P., Watts, S., McDonald, R. and Boucher, T. (2007) Domesticated nature: shaping landscapes and ecosystems for human welfare, *Science, 316*, 1866–1869.

Katz, E. (1992) The big lie: human restoration of nature. *Research in Philosophy and Technology, 12*, 231–241.

Katz, E. (2007) Book review of "Eric Higgs, *Nature by Design: People, Natural Process, and Ecological Restoration*," *Environmental Ethics, 29*, 213–216.

Keulartz, J. (2009) European nature conservation and restoration policy: problems and perspectives, *Restoration Ecology, 17*(4), 446–450.

Kiesecker, J.M., Copeland, H., Pocewicz, A. and McKenney, B. (2010) Development by design: blending landscape-level planning with the mitigation hierarchy, *Frontiers in Ecology and the Environment, 8*(5), 261–266.

Kleijn, D., Kohler, F., Báldi, A., Batáry, P., Concepción, E.D., Clough, Y., Díaz, M., Gabriel, D., Holzschuh, A., Knop, E., Kovács, A., Marshall, E.J., Tscharntke, T. and Verhulst, J. (2009) On the relationship between farmland biodiversity and land-use intensity in Europe. *Proceedings of the Royal Society B: Biological Sciences, 276*(1658), 903–909.

Kleinbauer, I., Dullinger, S., Peterseil, J. and Essl, F. (2010) Climate change might drive the invasive tree *Robinia pseudacacia* into nature reserves and endangered habitats, *Biological Conservation, 143*(2), 382–390.

Kortelainen, J. (2010) The European Green Belt: generating environmental governance – reshaping border areas, *Quaestiones Geographicae, 29*(4), 27–40.

Krebs, C. (2008) *The Ecological World View*, University of California Press, Berkeley, CA.

Kurtz, C. (2001) *A Practical Guide to Prairie Reconstruction*, Bur Oak Books, University of Iowa Press, Iowa City, IA.

Ladkin, D. (2005) Does "Restoration" necessarily imply the domination of nature?, *Environmental Values, 14*, 203–219.

Landres, P.B., Morgan, P. and Swanson, F.J. (1999) Overview of the use of natural variability concepts in managing ecological systems, *Ecological Applications, 9*(4), 1179–1188.

Latz, P. (2001) Landscape park Duisburg-Nord: the metamorphosis of an industrial site. In N. Kirkwood (ed.), *Manufactured Sites: Rethinking the Post-industrial Landscape*, Taylor and Francis, Abingdon, UK, pp. 150–161.

Lawler, J.J., Tear, T.H., Pyke, C., Shaw, M.R., Gonzalez, P., Kareiva, P., Hansen, L., Hannah, L., Klausmeyer, K., Aldous, A., Bienz, C. and Pearsall, S. (2010) Resource management in a changing and uncertain climate, *Frontiers in Ecology and the Environment, 8*(1), 35–43.

Leopold, A. (1939) The farmer as a conservationist, *American Forests, 45*, 294–316.

Leopold, A. (1949) *A Sand County Almanac and Sketches Here and There*, Oxford University Press, New York, NY.

Li, Y., Wang, W., Liu, Z. and Jiang, S. (2008) Grazing gradient versus restoration succession of *Leymus chinensis* (Trin.) Tzvel: grassland in Inner Mongolia, *Restoration Ecology, 16*(4), 572–583.

Light, A. (2000) Ecological restoration and the culture of nature: a pragmatic perspective. In P.H. Gobster and B. Hull (eds.), *Restoring Nature: Perspectives from the Social Sciences and Humanities*, Island Press, Washington, DC, pp. 49–70.

Light, A. (2006) Ecological citizenship: the democratic promise of restoration. In R.H. Platt (ed.), *The Humane Metropolis: People and Nature in the 21st Century City*, University of Massachusetts Press, Amherst, MA, pp. 169–181.

Ling, C., Handley, J. and Rodwell, J. (2003) Multifunctionality and scale in post-industrial land regeneration. In H.M. Moore, H.R. Fox and S. Elliott (eds.), *Land Reclamation: Extending the Boundaries*, A.A. Balkema Publishers, Lisse, Netherlands, pp. 27–34.

López-Hoffman, L., Varady, R.G., Flessa, K.W. and Balvanera, P. (2010) Ecosystem services across borders: a framework for transboundary conservation policy, *Frontiers in Ecology and the Environment, 8*(2), 84–91.

Lovejoy, T.E. (1985) Forest fragmentation in the Amazon: a case study. In H. Messel (ed.), *The Study of Populations*, Pergamon Press, Sydney, pp. 243–251.

Lovell, S.T. and Johnston, D.M. (2009) Creating multifunctional landscapes: how can the field of ecology inform the design of the landscape? *Frontiers in Ecology and the Environment, 7*, 212–220.

Lowenthal, D. (1985) *The Past is a Foreign Country*, Cambridge University Press, Cambridge, UK.

McCain, E.B. and Childs, J.L. (2008) Evidence of resident Jaguars (*Panthera Onca*) in the southwestern United States and the implications for conservation, *Journal of Mammalogy, 89*(1), 1–10.

McCann, J.M. (1999a) Before 1492, *Ecological Restoration, 17*(1), 15.

McCann, J.M. (1999b) Before 1492, *Ecological Restoration, 17*(3), 107.

McClain, R.J. and Lee, R.G. (1996) Adaptive management: promises and pitfalls, *Environmental Management, 20*(4), 437–448.

Mack, R.N., Simberloff, D., Lonsdale, W.M., Evans, H., Clout, M. and Bazzaz, F.A. (2000) Biotic invasions: causes, epidemiology, global consequences, and control. *Ecological Applications, 10*(3), 689–710.

McKay, J.K., Christian, C.E., Harrison, S. and Rice, K.J. (2005) "How local is local?": a review of practical and conceptual issues in the genetics of restoration, *Restoration Ecology, 13*(3), 432–440.

McKibben, B. (1989) *The End Of Nature*, Random House, New York, NY.

McKinney, M.L. and Lockwood, J.L. (1999) Biotic homogenization: a few winners replacing many losers in the next mass extinction, *Trends in Ecology and Evolution, 14*, 450–453.

McMenamin, S.K., Hadly, E.A. and Wright, C.K. (2008) Climatic change and wetland desiccation cause amphibian decline in Yellowstone National Park, *Proceedings of the National Academy of Sciences, 105*(44), 16988–16993.

Madgwick, J. and Jones, T.A. (2002) Europe. In M.R. Perrow and A.J. Davy (eds.), *Handbook of Ecological Restoration, Volume 2: Restoration in Practice*, Cambridge University Press, Cambridge, UK.

Majer, J.D., Brennan, K.E.C. and Moir, M.L. (2007) Invertebrates and the restoration of a forest ecosystem: 30 years of research following bauxite mining in Western Australia. *Restoration Ecology, 15*(suppl. 4), S104–S115.

Marris, E. (2009) Ragamuffin Earth, *Nature, 460*, 450–453.

Martinez, D. (2011, January 10) Slow death by carbon credits: editorial opinion, *The Boston Globe.* www.boston.com/bostonglobe/editorial_opinion/oped/articles/2011/01/10/slow_death_by_carbon_credits, retrieved April 15, 2011.

Maurer, B.A. (2006) Ecological restoration from a macroscopic perspective. In D.A. Falk, M.A. Palmer and J.B. Zedler (eds.), *Foundations of Restoration Ecology.* Island Press, Washington, DC, pp. 303–314.

Meekison, L. and Higgs, E. (1998) The rites of spring (and other seasons): the ritualizing of restoration, *Restoration and Management Notes, 16*, 73–81.

Memmott, J., Craze, P.G., Waser, N.M. and Price, M.V. (2007) Global warming and the disruption of plant–pollinator interactions, *Ecology Letters, 10*(8), 710–717.

Meyer, J.L., Frumhoff, P.C., Hamburg, S.P. and de la Rosa, C. (2010) Above the din but in the fray: environmental scientists as effective advocates, *Frontiers in Ecology and the Environment, 8*(6), 299–305.

Meyer, M.J., Crawford, J.A. and Allison, S.K. (2002) Geographic distribution: *Hemidactylium scutatum, Herpetological Review, 33*, 217.

Michael, M.A. (2001) How to interfere with nature, *Environmental Ethics, 23*, 135–154.

Millennium Ecosystem Assessment (2005) *Ecosystems and Human Well-being: Synthesis*, Island Press, Washington, DC.

Miller, J.R. (2005) Biodiversity conservation and the extinction of experience, *Trends in Ecology & Evolution, 20*(8), 430–434.

Miller-Rushing, A. and Primack, R.B. (2008) Global warming and flowering times in Thoreau's Concord: a community perspective, *Ecology, 89*(2), 332–341.

Minteer, B.A. and Collins, J.P. (2010) Move it or lose it? The ecological ethics of relocating species under climate change. *Ecological Applications, 20*(7), 1801–1804.

Mitchell, J.H. (2001) *The Wildest Place on Earth: Italian Gardens and the Invention of Wilderness*, Counterpoint Press, New York, NY.

Mooney, H.A. (2010) The ecosystem-service chain and the biological diversity crisis, *Philosophical Transactions of the Royal Society B: Biological Sciences, 365*(1537), 31–39.

Moore, S.A., Wallington, T.J., Hobbs, R.J., Ehrlich, P.R., Holling, C.S., Levin, S., Lindenmayer, D., Pahl-Wostl, C., Possingham, H., Turner, M.G. and Westoby, M. (2009) Diversity in current ecological thinking: implications for environmental management. *Environmental Management, 43*, 17–27.

Munro, J.W. (2006) Ecological restoration and other conservation practices: the difference, *Ecological Restoration, 24*(3), 182–189.

Naeem, S. (2006) Biodiversity and ecosystem functioning in restored ecosystems: extracting principles for a synthetic perspective. In D.A. Falk, M.A. Palmer and J.B. Zedler (eds.), *Foundations of Restoration Ecology*, Island Press, Washington, DC, pp. 210–237.

Nellemann, C. and Corcoran, E. (eds.) (2010) *Dead Planet, Living Planet: Biodiversity and Ecosystem Restoration for Sustainable Development – A Rapid Response Assessment.* United Nations Environment Programme, Nairobi, Kenya. www.grida.no/publications/rr/dead-planet.

Nelson, E., Mendoza, G., Regetz, J., Polasky, S., Tallis, H., Cameron, D.R., Chan, K.M.A., Daily, G.C., Goldstein, J., Kareiva, P.M., Lonsdorf, E., Naidoo, R., Ricketts, T.H. and Shaw, M.R. (2009) Modeling multiple ecosystem services, biodiversity conservation, commodity production, and tradeoffs at landscape scales. *Frontiers in Ecology and the Environment, 7*(1), 4–11.

Newmark, W.D. (1986) Species–area relationship and its determinants for mammals in western North American national parks, *Biological Journal of the Linnean Society, 28*(1–2), 83–98.

Norton, D.A. (2009) Species invasions and the limits to restoration: learning from the New Zealand experience, *Science, 325*(5940), 569–571.

Oliveira-Santos, L.G.R. and Fernandez, F.A.S. (2010) Pleistocene rewilding, Frankenstein ecosystems, and an alternative conservation agenda, *Conservation Biology, 24*(1), 4–5.

Olson, R. (2009) *Don't be Such a Scientist: Talking Substance in an Age of Style*, Island Press, Washington, DC.

Olwig, K. (1995) Reinventing common nature: Yosemite and Mount Rushmore – a meandering tale of a double nature. In W. Cronon (ed.), *Uncommon Ground: Rethinking the Human Place in Nature*, W.W. Norton Company, New York, NY, pp. 379–408.

Packard, S. (1988) Chronicles of restoration: just a few oddball species – restoration and the rediscovery of the tallgrass savanna. *Restoration Management Notes, 6*(1), 13–22.

Paine, R.T. (1966) Food web complexity and species diversity, *American Naturalist, 100*, 65–75.

Pandey, D.N. (2002) Sustainability science for mine-spoil restoration, *Current Science, 83*, 792–793.

Parmesan, C. (2006) Ecological and evolutionary responses to recent climate change. *Annual Review of Ecology, Evolution and Systematics, 37*, 637–669.

Pauchard, A., Kueffer, C., Dietz, H., Daehler, C.C., Alexander, J., Edwards, P.J., Arévalo, J.R., Cavieres, L.A., Guisan, A., Haider, S., Jakobs, G., McDougall, K., Millar, C.I., Naylor, B.J., Parks, C.G., Rew, L.J. and Seipel, T. (2009) Ain't no mountain high enough: plant invasions reaching new elevations. *Frontiers in Ecology and the Environment, 7*(9), 479–486.

Peng, S., Hou, Y. and Chen, B. (2009) Vegetation restoration and its effects on carbon balance in Guangdong Province, China, *Restoration Ecology, 17*(4), 487–494.

Prach, K., Bartha, S., Joyce, C.B., Pysek, P., van Diggelen, R. and Wiegleb, G. (2001a) The role of spontaneous vegetation succession in ecosystem restoration: a perspective, *Applied Vegetation Science, 4*(1), 111–114.

Prach, K., Pysek, P. and Bastl, M. (2001b) Spontaneous vegetation succession in human-disturbed habitats: a pattern across seres. *Applied Vegetation Science, 4*(1), 83–88.

Pyne, S.J. (1982) *Fire in America: A Cultural History of Wildland and Rural Fire*, Princeton University Press, Princeton, NJ.

Rackham, O. (1986) *The History of the Countryside*, J.M. Dent, London.

Rands, M.R.W., Adams, W.M., Bennun, L., Butchart, S.H.M., Clements, A., Coomes, D., Entwistle, A., Hodge, I., Kapos, V., Scharlemann, J.P.W., Sutherland, W.J. and Vira, B. (2010) Biodiversity conservation: challenges beyond 2010, *Science, 329*(5997), 1298–1303.

Ravenscroft, C., Scheller, R.M., Mladenoff, D.J. and White, M.A. (2010) Forest restoration in a mixed-ownership landscape under climate change, *Ecological Applications, 20*(2), 327–346.

Rayfield, B., Anand, M. and Laurence, S. (2005) Assessing simple versus complex restoration strategies for industrially disturbed forests. *Restoration Ecology, 13*(4), 639–650.

Rees, W.E. (2000) Patch disturbance, ecofootprints, and biological integrity: revisiting the limits to growth (or why industrial society is inherently unsustainable). In D. Pimentel, L. Westra and R.F. Noss (eds.), *Ecological Integrity: Integrating Environment, Conservation, and Health,* Island Press, Washington, DC, pp. 139–158.

Rey Benayas, J.M., Newton, A.C., Diaz, A. and Bullock, J.M. (2009) Enhancement of biodiversity and ecosystem services by ecological restoration: a meta-analysis, *Science, 325*(5944), 1121–1124.

Ricciardi, A. (2007) Are modern biological invasions an unprecedented form of global change?, *Conservation Biology, 21*(2), 329–336.

Ricciardi, A. and Simberloff, D. (2009) Assisted colonization is not a viable conservation strategy, *Trends in Ecology and Evolution, 24*(5), 248–253.

Richardson, D.M., Hellmann, J.J., McLachlan, J.S., Sax, D.F., Schwartz, M.W., Gonzalez, P., Brennan, E.J., Camacho, A., Root, T.L., Sala, O.E., Schneider, S.H., Ashe, D.M., Clark, J.R., Early, R., Etterson, J.R., Fielder, E.D., Gill, J.L., Minteer, B.A., Polasky, S., Safford, H.D., Thompson, A.R. and Vellend, M. (2009) Multidimensional evaluation of managed relocation, *Proceedings of the National Academy of Sciences, 106*(24), 9721–9724.

Ridder, B. (2007) The naturalness versus wildness debate: ambiguity, inconsistency, and unattainable objectivity, *Restoration Ecology, 15*(1), 8–12.

Rodwell, J. and Skelcher, G. (2002) *The Ecoscapes and Plant Communities of Cheshire,* University of Lancaster Unit of Vegetation Science, Lancaster, UK.

Rosenzweig, M.L. (2003) *Win–Win Ecology: How the Earth's Species Can Survive in the Midst of Human Enterprise,* Oxford University Press, New York, NY.

Rowe, H.I. (2010) Tricks of the trade: techniques and opinions from 38 experts in tallgrass prairie restoration, *Restoration Ecology, 18,* 253–262.

Rubenstein, D.R., Rubenstein, D.I., Sherman, P.W. and Gavin, T.A. (2006) Pleistocene Park: does re-wilding North America represent sound conservation for the 21st century? *Biological Conservation, 132*(2), 232–238.

Ruiz-Jaen, M.C. and Aide, T.M. (2005) Restoration success: how is it being measured? *Restoration Ecology, 13,* 569–577.

Sagoff, M. (1983) At the shrine of Our Lady of Fatima, or, why political questions are not all economic. In D. Scherer and T. Attig (eds.), *Ethics and the Environment,* Prentice-Hall, Englewood Cliffs, NJ, pp. 221–234.

Sagoff, M. (2009) The economic value of ecosystem services, *Bioscience, 59*(6), 461.

Sanderson, E.W., Jaiteh, M., Levy, M.A., Redford, K.H., Wannebo, A.V. and Woolmer, G. (2002) The human footprint and the last of the wild. *Bioscience, 52*(10), 891–904.

Saxon, E., Baker, B., Hargrove, W., Hoffman, F. and Zganjar, C. (2005) Mapping environments at risk under different global climate change scenarios, *Ecology Letters, 8*(1), 53–60.

Schramm, P. (1992) Prairie restoration: a twenty-five year perspective on establishment and management. In D.D. Smith and C.A. Jacobs (eds.), *Proceedings of the*

Twelfth North American Prairie Conference: Recapturing a Vanishing Heritage, University of Northern Iowa Press, Cedar Falls, IA, pp. 169–177.

Scott, R. and Whitbread-Abrutat, P. (2010) Eyes wide open: building bridges and crossing them (Workshop presented at the European Chapter of the Society of Ecological Restoration meeting, Avignon, France, August 23–27, 2010).

Seastedt, T.R., Hobbs, R.J. and Suding, K.N. (2008) Management of novel ecosystems: are novel approaches required? *Frontiers in Ecology and the Environment, 6*(10), 547–553.

Seddon, P.J. (2010) From reintroduction to assisted colonization: moving along the conservation translocation spectrum, *Restoration Ecology, 18*(6), 796–802.

Shanley, P. and López, C. (2009) Out of the loop: why research rarely reaches policy makers and the public and what can be done, *Biotropica, 41*(5), 535–544.

Slootweg, R. and van Beukering, P. (2008) *Valuation of Ecosystem Services and Strategic Environmental Assessment: Lessons from Influential Cases,* Commission for Environmental Assessment, Utrecht, the Netherlands.

Smits, N.A.C., Willems, J.H. and Bobbink, R. (2008) Long-term after-effects of fertilisation on the restoration of calcareous grasslands. *Applied Vegetation Science, 11*(2), 279–286.

Snook, S.A. (2002) *Friendly Fire: The Accidental Shootdown of US Black Hawks Over Northern Iraq,* Princeton University Press, Princeton, NJ.

Society for Ecological Restoration Science and Policy Working Group (2004) *The SER Primer on Ecological Restoration.* www.ser.org, retrieved July 1, 2007.

Soulé, M. (1986) Conservation biology and the "real world." In M. Soulé (ed.), *Conservation Biology: The Science of Scarcity and Diversity,* Sinauer Associates Inc., Sunderland, MA, pp. 1–12.

Stevens, W.K. (1995) *Miracle Under the Oaks: The Revival of Nature in America,* Pocket Books, New York, NY.

Swart, J.A.A. and van der Windt, H.J. (2010) Boundary work in ecological restoration and conservation, *Seventh Society for Ecological Restoration European Conference on Ecological Restoration,* Avignon, France, August 23–27.

Temperton, V.M., Hobbs, R.J., Nuttle, T. and Halle, S. (2004) *Assembley Rules and Restoration Ecology,* Island Press, Washington, DC.

Thompson, S., Larcom, A. and Lee, J.T. (1999) Restoring and enhancing rare and threatened habitats under agri-environment agreements: a case study of the Chiltern Hills area of outstanding natural beauty, UK, *Land Use Policy, 16*(2), 93–105.

Thuller, W. (2007) Climate change and the ecologist, *Nature, 448,* 550–552.

Tischew, S. and Kirmer, A. (2007) Implementation of basic studies in the ecological restoration of surface-mined land, *Restoration Ecology, 15*(2), 321–325.

Tomblin, D.C. (2009) The ecological restoration movement: diverse cultures of practice and place, *Organization and Environment, 22,* 185–207.

Turner, F. (1985) Cultivating the American garden: toward a secular view of nature, *Harper's,* August, 45–52.

Turner, F. (1987) The humble bee: restoration as natural reproduction, *Restoration and Management Notes, 5*(1), 15–17.

UNFCCC (2010) Decision 4/CP.15, Methodological guidance for activities relating to reducing emissions from deforestation and forest degradation and the role of conservation, sustainable management of forest and enhancement of forest carbon stocks in developing countries. UNFCCC, Bonn, Germany. http://

unfccc.int/resource/docs/2009/cop15/eng/11a01.pdf#page=11, retrieved April 27, 2011.

UN-REDD. (2008) *UN Collaborative Programme on Reducing Emissions from Deforestation and Forest Degradation in Developing Countries (UN-REDD).* United Nations Environment Program.

USDA (2011) Conservation reserve program celebrates 25 years. www.fsa.usda.gov/FSA/printapp?Name=nr_20101223_rel0671.html&newstype=newsrel, retrieved May 3, 2011.

Vale, T.R. (ed.) (2002) *Fire, Native Peoples, and the Natural Landscape,* Island Press, Washington, DC.

van Andel, J. and Aronson, J. (eds.) (2006) *Restoration Ecology,* Blackwell Publishing, Malden, MA.

van der Heijden, H.-A. (2005) Ecological restoration, environmentalism and the Dutch politics of "New Nature," *Environmental Values, 14,* 427–446.

van der Veken, S., Hermy, M., Vellend, M., Knapen, A. and Verheyen, K. (2008) Garden plants get a head start on climate change, *Frontiers in Ecology and the Environment, 6*(4), 212–216.

Vidra, R.L. (2006) Studying the ethics of ecological restoration: an introduction, *Ecological Restoration, 24*(2), 100–101.

Vidra, R.L. and Shear, T.H. (2010) Ethical dimensions of ecological restoration. In F.A. Comin (ed.), *Ecological Restoration: A Global Challenge,* Cambridge University Press, Cambridge, UK, pp. 100–111.

Vilà, M., Basnou, C., Pyšek, P., Josefsson, M., Genovesi, P., Gollasch, S., Nentwig, W., Olenin, S., Roques, A., Roy, D. and Hulme, P.E. (2010) How well do we understand the impacts of alien species on ecosystem services? A pan-European, cross-taxa assessment, *Frontiers in Ecology and the Environment, 8*(3), 135–144.

Vitousek, P.M., Ehrlich, P.R., Ehrlich, A.H. and Matson, P.A. (1986) Human appropriation of the products of photosynthesis, *Bioscience, 36*(6), 368–373.

Vitousek, P.M., Mooney, H.A., Lubchenco, J. and Melillo, J.M. (1997) Human domination of Earth's ecosystems, *Science, 277,* 494–499.

Vogel, S. (2003) The nature of artifacts, *Environmental Ethics, 25,* 149–168.

Waldman, J. (2008) With temperatures rising, here comes "global weirding." *Yale: Environment 360.* http://e360.yale.edu/content/print.msp?id=2132, retrieved March 25, 2009.

Watson, A. (2009) Final warning from a sceptical prophet, *Nature, 458,* 970–971.

Webb, H. (2010) How long does it take to make a meadow?. www.nude-ewe.com/blog/?p=871, retrieved April 27, 2011.

White, L.T. (1967) The historical roots of our ecologic crisis, *Science, 155,* 1203–1207.

Whited, T.L. (1996) Alpine myths and metaphors: the debate over restoration in nineteenth century France, *Restoration and Management Notes, 14,* 53–56.

Wilkinson, D.M. (2004) Guest editorial: the parable of Green Mountain – Ascension Island, ecosystem construction and ecological fitting, *Journal of Biogeography, 31,* 1–4.

Willems, J.H. (2001) Problems, approaches and results in restoration of Dutch calcareous grassland during the last 30 years. *Restoration Ecology, 9*(2), 147–154.

Willis, C.G., Ruhfel, B., Primack, R.B., Miller-Rushing, A.J. and Davis, C.C. (2008) Phylogenetic patterns of species loss in Thoreau's woods are driven by climate change, *Proceedings of the National Academy of Sciences, 105*(44), 17029–17033.

Wilson, E.O. (1984) *Biophilia*, Harvard University Press, Cambridge, MA.

Winterhalder, K. (1984) Environmental degradation and rehabilitation in the Sudbury area. *Laurentian University Reviews, 16*, 15–47.

Winterhalder, K., Clewell, A.F. and Aronson, J. (2004) Values and science in ecological restoration: a response to Davis and Slobodkin, *Restoration Ecology, 12*, 4–7.

Woodcock, B.A., Pywell, R.F., Roy, D.B., Rose, R.J. and Bell, D. (2005) Grazing management of calcareous grasslands and its implications for the conservation of beetle communities, *Biological Conservation, 125*(2), 193–202.

Woodley, S. (2010) Ecological integrity: a framework for ecosystem-based management. In D.N. Cole and L. Yung (eds.), *Beyond Naturalness: Rethinking Park And Wilderness Stewardship in an Era of Rapid Change*, Island Press, Washington, DC, pp. 106–124.

Woodworth, P. (2010) A personal assessment of the SERI conference in Perth and what it reveals about the Society, *Ecological Restoration, 28*(1), 4–6.

Wu, L. (2004) Review of 15 years of research on ecotoxicology and remediation of land contaminated by agricultural drainage sediment rich in selenium. *Ecotoxicology and Environmental Safety, 57*(3), 257–269.

Xu, Z., Wan, S., Zhu, G., Ren, H. and Han, X. (2010) The influence of historical land use and water availability on grassland restoration, *Restoration Ecology, 18*, 217–225.

Yin, R. and Yin, G. (2010) China's primary programs of terrestrial ecosystem restoration: initiation, implementation, and challenges, *Environmental Management, 45*(3), 429–441.

Yin, R., Yin, G. and Li, L. (2010) Assessing China's ecological restoration programs: what's been done and what remains to be done? *Environmental Management, 45*(3), 442–453.

Young, T.P. (2000) Restoration ecology and conservation biology, *Biological Conservation, 92*(1), 73–83.

Young, T.P., Petersen, D.A. and Clary, J.J. (2005) The ecology of restoration: historical links, emerging issues and unexplored realms. *Ecology Letters, 8*, 662–673.

Zar, J.H. (1999) *Biostatistical Analysis*, 4th edn., Prentice-Hall, Upper Saddle River, NJ.

Zhou, P., Luukkanen, O., Tokola, T. and Nieminen, J. (2008) Vegetation dynamics and forest landscape restoration in the Upper Min River watershed, Sichuan, China, *Restoration Ecology, 16*(2), 348–358.

Index

Page numbers in **bold** denote figures.

Printed in the United States
by Baker & Taylor Publisher Services

Printed in the United States
by Baker & Taylor Publisher Services